# Microbiological Corrosion of Buildings

# Occupational Safety, Health, and Ergonomics: Theory and Practice

***Series Editor: Danuta Koradecka***

***(Central Institute for Labour Protection – National Research Institute)***

This series will contain monographs, references, and professional books on a compendium of knowledge in the interdisciplinary area of environmental engineering, which covers ergonomics and safety and the protection of human health in the working environment. Its aim consists in an interdisciplinary, comprehensive and modern approach to hazards, not only those already present in the working environment, but also those related to the expected changes in new technologies and work organizations. The series aims to acquaint both researchers and practitioners with the latest research in occupational safety and ergonomics. The public, who want to improve their own or their family's safety, and the protection of heath will find it helpful, too. Thus, individual books in this series present both a scientific approach to problems and suggest practical solutions; they are offered in response to the actual needs of companies, enterprises, and institutions.

**Individual and Occupational Determinants: Work Ability in People with Health Problems**
*Joanna Bugajska, Teresa Makowiec-Dąbrowska, Tomasz Kostka*

**Healthy Worker and Healthy Organization: A Resource-Based Approach**
*Dorota Żołnierczyk-Zreda*

**Emotional Labour in Work with Patients and Clients: Effects and Recommendations for Recovery**
*Dorota Żołnierczyk-Zreda*

**New Opportunities and Challenges in Occupational Safety and Health Management**
*Daniel Podgórski*

**Emerging Chemical Risks in the Work Environment**
*Małgorzata Pośniak*

**Visual and Non-Visual Effects of Light: Working Environment and Well-Being**
*Agnieszka Wolska, Dariusz Sawicki, Małgorzata Tafil-Klawe*

**Occupational Noise and Workplace Acoustics: Advances in Measurement and Assessment Techniques**
*Dariusz Pleban*

**Virtual Reality and Virtual Environments: A Tool for Improving Occupational Safety and Health**
*Andrzej Grabowski*

**Head, Eye, and Face Personal Protective Equipment: New Trends, Practice and Applications**
*Katarzyna Majchrzycka*

**Nanoaerosols, Air Filtering and Respiratory Protection: Science and Practice**
*Katarzyna Majchrzycka*

**Microbiological Corrosion of Buildings: A Guide to Detection, Health Hazards and Mitigation**
*Rafał L. Górny*

**Respiratory Protection Against Hazardous Biological Agents**
*Katarzyna Majchrzycka, Justyna Szulc, Małgorzata Okrasa*

For more information about this series, please visit: https://www.crcpress.com/Occupational-Safety-Health-and-Ergonomics-Theory-and-Practice/book-series/CRCOSHETP

# Microbiological Corrosion of Buildings

## A Guide to Detection, Health Hazards and Mitigation

Edited by
Rafał L. Górny

CRC Press is an imprint of the
Taylor & Francis Group, an informa business

First edition published 2021
by CRC Press
6000 Broken Sound Parkway NW, Suite 300, Boca Raton, FL 33487-2742

and by CRC Press
2 Park Square, Milton Park, Abingdon, Oxon, OX14 4RN



CRC Press is an imprint of Taylor & Francis Group, LLC



ISBN: 978-0-367-49984-6 (hbk)
ISBN: 978-0-367-49989-1 (pbk)
ISBN: 978-1-003-04843-5 (ebk)

Typeset in Times
by Deanta Global Publishing Services, Chennai, India

# Contents

# Preface

Buildings are constantly exposed to colonisation by microorganisms. This type of adverse impact is particularly evident when their specific construction details are subject to environmental stress caused by the presence of water. Such a situation is especially noticeable in the case of both minor (dampness, swamping) and major (flooding) damage caused by water. The publishing market still lacks elaborations which, in a professional and reader-friendly way, combine purely technical issues related to the destruction of buildings caused by water in its various forms with the issues regarding microbiological contamination of interiors. The resulting questions concerning the identification and assessment of microbiological contamination of indoor spaces, as well as health and epidemiological consequences of exposure to microorganisms in buildings which have undergone microbiological corrosion have yet to be answered, until now.

The following monograph, titled *Microbiological Corrosion of Buildings: A Guide to Detection, Health Hazards and Mitigation*, addresses the above-mentioned issues in a holistic way. The contributing authors describe historical and contemporary problems related to microbiological contamination of buildings, characterise microorganisms, structures and substances originating from them which are responsible for the corrosion of buildings and result from it, as well as adverse health effects on people in the buildings. They also discuss the environmental factors favouring microbiological contamination of buildings, most frequently observed symptoms of biodeterioration of construction and finishing materials and technical methods of drying buildings, as well as physical and chemical methods of combating bio-corrosion in them. They also acquaint the reader with the methods of identification and assessment of indoor microbiological hazards and the scope of activities carried out to check the effectiveness of remediation measures. What may also be of interest to readers is that the monograph contains practical examples of bio-corrosion in residential, office, industrial, agricultural, healthcare, historic and special buildings, including libraries and archives.

By giving the following book to our readers, we hope it will be well received by engineers and technicians who are professionally involved in the construction sector or the protection of buildings against corrosion, as well as by manufacturers of construction and finishing materials, architects and interior designers, and scientists who are active in the fields of construction, material engineering, biology, medicine or public health and who are interested in these issues as part of their professional activities. We also hope that the issues described in the book will be of interest to a large number of regular users, i.e. people who live and work in different types of buildings and are not indifferent to their technical and sanitary conditions.

**Rafał L. Górny**

# Series Editor

**Professor Danuta Koradecka, PhD, D.Med.Sc.** and Director of the Central Institute for Labour Protection – National Research Institute (CIOP-PIB), is a specialist in occupational health. Her research interests include the human health effects of hand-transmitted vibration; ergonomics research on the human body's response to the combined effects of vibration, noise, low temperature and static load; assessment of static and dynamic physical load; development of hygienic standards as well as development and implementation of ergonomic solutions to improve working conditions in accordance with International Labour Organisation (ILO) convention and European Union (EU) directives. She is the author of more than 200 scientific publications and several books on occupational safety and health.

***

The "Occupational Safety, Health, and Ergonomics: Theory and Practice" series of monographs is focused on the challenges of the 21st century in this area of knowledge. These challenges address diverse risks in the working environment of chemical (including carcinogens, mutagens, endocrine agents), biological (bacteria, viruses), physical (noise, electromagnetic radiation) and psychophysical (stress) nature. Humans have been in contact with all these risks for thousands of years. Initially, their intensity was lower, but over time it has gradually increased, and now too often exceeds the limits of man's ability to adapt. Moreover, risks to human safety and health, so far assigned to the working environment, are now also increasingly emerging in the living environment. With the globalisation of production and merging of labour markets, the practical use of the knowledge on occupational safety, health, and ergonomics should be comparable between countries. The presented series will contribute to this process.

The Central Institute for Labour Protection – National Research Institute, conducting research in the discipline of environmental engineering, in the area of working environment and implementing its results, has summarised the achievements – including its own – in this field from 2011 to 2019. Such work would not be possible without cooperation with scientists from other Polish and foreign institutions as authors or reviewers of this series. I would like to express my gratitude to all of them for their work.

It would not be feasible to publish this series without the professionalism of the specialists from the Publishing Division, the Centre for Scientific Information and Documentation, and the International Cooperation Division of our Institute. The challenge was also the editorial compilation of the series and ensuring the efficiency of this publishing process, for which I would like to thank the entire editorial team of CRC Press – Taylor & Francis Group.

***

This monograph, published in 2020, has been based on the results of a research task carried out within the scope of the second to fourth stage of the Polish National

Programme "Improvement of safety and working conditions" partly supported – within the scope of research and development – by the Ministry of Science and Higher Education/National Centre for Research and Development, and within the scope of state services – by the Ministry of Family, Labour and Social Policy. The Central Institute for Labour Protection – National Research Institute is the Programme's main coordinator and contractor.

# Editor

**Rafał L. Górny, PhD**, received his medical education (MS degree in 1992, PhD in 1999, DSci ['habilitation'] degree in 2006, Full Professor of medical sciences in 2014) at Medical University of Silesia (formerly Silesian Medical Academy), Katowice, Poland. In the past 25 years of his professional career, he has been engaged in numerous studies devoted to health-related aspects of exposure to particulate (including biological) and fibrous aerosols in occupational and non-occupational environments. His research efforts have been presented in more than 70 peer-reviewed publications, more than 100 conference presentations and several monographs and book chapters. He is currently head of the Laboratory of Biohazards at the Central Institute for Labour Protection–National Research Institute, Warsaw, Poland. He is also head of the Biological Agents' Expert Group at the Interdepartmental Commission for Maximum Admissible Concentrations and Intensities for Agents Harmful to Health in the Working Environment, Warsaw, Poland. Since 2002, he has been serving as an adviser to World Health Organization and European Commission within the field of biological agents.

# Contributors

**Marcin Cyprowski, PhD in Medical Sciences**, is a research assistant in the Laboratory of Biohazards at the Central Institute for Labour Protection – National Research Institute, Warsaw, Poland. For over 19 years he has dealt with the problems of biological hazards in the working and living environments, especially among employees of waste management, sewage treatment plants, metal-finishing plants, pig and poultry farms, as well as in various public buildings (kindergartens, offices) and in dwellings. He has authored or co-authored more than 35 peer-reviewed scientific publications, and participated in numerous scientific conferences. His professional interest is the assessment of toxic effects of bioaerosols on workers' health.

**Małgorzata Gołofit-Szymczak, PhD in Environmental Engineering**, is a research assistant in the Laboratory of Biohazards at the Central Institute for Labour Protection – National Research Institute, Warsaw, Poland. She has over 24 years of research experience in studying microbiological pollution in occupational and non-occupational environments, pollution of air-conditioning systems and risk assessment posed by microbiological agents. Experienced in teaching and development of training programs and materials in the field of harmful biological agents, Dr Gołofit-Szymczak has published over 30 articles in internationally peer-reviewed journals.

**Anna Ławniczek-Wałczyk, PhD in Environmental Engineering**, is a research assistant in the Laboratory of Biohazards at the Central Institute for Labour Protection – National Research Institute, Warsaw, Poland. For over 12 years, her research activities have focused on occupational exposure to biological hazards (especially bioaerosols) including the methods of their quantification, the assessment of the relationship between work exposure and health outcomes, monitoring the transmission of airborne pathogens in human communities using genotypic methods and understanding the basis of their resistance to antibiotics and disinfectants. She has authored or co-authored more than 35 peer-reviewed scientific publications and training materials, and participated in numerous conferences. She is also actively involved in teaching activities.

**Agata Stobnicka-Kupiec, PhD (Eng) in Agricultural Sciences**, is a research assistant in the Laboratory of Biohazards at the Central Institute for Labour Protection – National Research Institute, Warsaw, Poland. Since 2013, she has been engaged in studies regarding biological contamination of occupational and non-occupational environments. She is the author or co-author of 34 peer-reviewed publications and 20 conference presentations. She has been working in the field of the qualitative and quantitative assessment of microbial contamination of indoor spaces. Her professional interest is the application of molecular methods in the detection of viruses in bioaerosols and on fomites.

# Glossary

**Absorption:** process whereby gas or vapour molecules are transferred to the liquid or solid phase or liquid substance is transferred to the solid phase

**Actinomycetes:** filamentous Gram-positive, aerobic or anaerobic bacteria belonging to the *Actinobacteria* phylum

**Adsorption:** transfer of gas or vapour molecules from the surrounding gas to the liquid or solid surface or liquid substance to a solid surface

**Aerosol:** assembly of liquid or solid particles suspended in a gaseous medium

**Allergen:** substance that can cause an allergic reaction in a sensitised person

**Allergic alveolitis:** group of respiratory diseases caused by repeated inhalation exposure to organic dusts with subsequent sensitisation to their components

**Allergic bronchopulmonary aspergillosis (ABPA):** pulmonary disorder caused by hypersensitivity to fungal antigens of *Aspergillus* genera, most often of *A. fumigatus* species

**Allergy:** adverse health reaction upon secondary contact with the antigen

**Antibody:** immunoglobulin, a protein in the blood produced by B lymphocytes, which has the ability to specifically recognise antigens; the following immunoglobulins are distinguished: IgA, IgD, IgE, IgG and IgM

**Antigen:** typically a protein that results in the formation of antibodies; a molecule which reacts with the antibody through specific receptors on T and B lymphocytes

**Archaea:** tiny single-celled organisms, which lack cell nuclei and are usually extremophile

**Atopy:** the inherited predisposition to develop allergic reactions as a result of increased levels of specific immune globulin E (IgE) antibodies that target sensitising agents or allergens

**Bacillus:** a rod-shaped bacterium

**Bacteria:** large group of prokaryotic microorganisms with one chromosome in a nuclear region and which only replicate asexually through cell division

**Bioaerosol:** particles of biological origin suspended in a gaseous medium (e.g. in the air)

**Biocidal product:** active substance or preparation containing at least one or more active substances, intended to destroy, deter, render harmless, prevent the action of, or otherwise exert a controlling effect on any harmful organism by chemical or biological means

**Bio-corrosion (microbiological corrosion):** corrosion occurring under the influence of microorganisms (mainly bacteria and fungi) and their metabolites

**Biodeterioration:** process of microbiological decomposition; undesirable phenomenon caused by microbiological agents

**Biofilm:** complex multicellular structure of microbes and other organisms, which is surrounded by a layer of organic and inorganic substances produced by these microbes, exhibiting adhesion to the surface on which it forms

**Byssinosis:** a respiratory disease caused by exposure to cotton, flax and/or hemp dust

**Carcinogenic:** substance capable of causing cancer

**Carcinogenicity:** ability to cause cancer or increase the likelihood of its occurrence

**Cell membrane:** a semi-permeable biological membrane separating the inside of a cell from the outside world

**Cell wall:** the outermost layer of bacterial and fungal cells

**Cellulolytic:** capable of hydrolysis of cellulose molecules and using its decomposition products as a source of carbon and energy

**Coccus:** any bacterium or archaeon that has a spherical, ovoid or generally round shape

**Colony forming unit (CFU):** unit by which the culturable number of microorganisms is given

**Condensation:** process in which more vapour molecules are arriving at a particle's surface than are leaving the surface, resulting in a net growth of the particle

**Conidium:** asexual, vegetative, non-motile propagule, not formed by cleavage

**Containment measures:** measures that are used to prevent or reduce the accidental transfer or release of a biological agent from its source

**Contamination:** action or state of making or being made impure by polluting or poisoning

**Culturable:** single microbial cells or their aggregates able to form colonies on a solid nutrient medium

**Cytokines:** soluble molecules which mediate intercellular communication

**Cytoplasm:** part of a cell (excluding nucleus) containing permanent structural elements

**Dalton (Da):** conventional relative atomic mass unit; one Dalton equals 1/12 of the mass of an atom of carbon C12 isotope: $1\ \text{Da} = 1.66 \times 10^{-24}$ g

**Dehydration:** removal of water particles

**Denaturation:** disruption of hydrogen bonds leading to the loss of biological activity

**Deoxyribonucleic acid (DNA):** macromolecular organic chemical compound belonging to nucleic acids; it is found in chromosomes and acts as a carrier of genetic information of living organisms

**Dew point (also dew point temperature):** temperature to which the air would have to cool (at constant pressure and water vapour content) in order to reach saturation

**Disinfection:** process aimed at minimising the number of microorganisms with a chemical on inanimate objects

**Ecosystem:** the complex of living organisms, their physical environment and all their interrelationships in a particular unit of space

**Endotoxin:** constituent of the outer membrane of Gram-negative bacteria (lipopolysaccharide), consisting of a complex lipid, lipid A, which is covalently bound to a polysaccharide
**Endotoxin unit (EU):** unit standardized against the defined reference material (reference standard endotoxin)
**Enzyme-linked immunosorbent assay (ELISA):** assay in which an enzyme is linked to an antibody and a coloured substrate is used to measure the activity of bound enzyme and, hence, the amount of bound antibody
**Enzymes:** macromolecular, mostly protein compounds that catalyse chemical reactions
**Eosinophil:** granulocytes of the immune system cells, which play an essential role in combating parasites and allergic reactions
**Epitope (antigenic determinant):** part of the antigen that directly binds to an antibody
**Equilibrium relative humidity (ERH):** the water activity of a material expressed as a percentage
**Extremophile:** organism which tolerates or requires extreme variability of environmental factors in order to live.
**Fibrillation:** art conservatory method of supplementing gaps and consolidation of fabrics with a natural fibre mass
**Filtration:** collection of particles suspended in gas or liquid by flowing through a porous medium
**Fungi:** diverse group of eukaryotic microorganisms with membrane-bound nucleus comprising several chromosomes
**Germination:** process leading to activation of spore after a period of dormancy
**Glucan:** polysaccharide molecule present in the cell walls of eukaryotes and prokaryotes including most moulds, upper fungi, yeasts, algae and certain bacteria
**Glycoproteins:** proteins containing, usually numerous, oligosaccharides (several saccharide units) covalently bound to them
**Gram-negative:** not retaining the primary stain (crystal violet) during the Gram staining procedure.
**Gram-positive:** retaining the primary stain (crystal violet) during the Gram staining procedure
**Granulocytes:** category of leukocytes, which have numerous granules in their cytoplasm and a cell nucleus divided into segments
**Green building:** human activity related to the design and construction of environmentally friendly and human friendly buildings
**Hazard:** probability that a particular danger (threat) occurs within a given period of time
**Heterogeneous:** consisting of individual components that may differ from each other in size, shape and chemical composition
**Homogeneous:** consisting of individual components of the same or a similar kind or nature
**Hydrophilic:** having a strong affinity for water for growth; growing under conditions of high water availability

**Hydrophilicity:** tendency of chemical particles to bond with water
**Hydrophobicity:** tendency of chemical particles to repel water molecules
**Hypersensitivity pneumonitis (HP):** group of immunologically mediated, granulomatous lung diseases caused by repeated inhalation and sensitisation to any of wide array of organic agents including microbial ones, animal proteins, and low-molecular-weight chemical compounds
**Hyphae:** (vegetative) filament of mycelium, without or with cross-walls
**Immunomodulator:** substance that affects the immune system
**Immunotoxicity:** over- or under-activation of the immune system by a factor(s) featuring high biological activity
**Impaction:** collection of airborne particles accelerated through the nozzle or orifice on a surface by the inertia effect
**Impingement:** combination of impaction onto a surface and subsequent dispersion into a liquid medium
**Inhibitor:** chemical compound that inhibits or slows down a chemical reaction
**Injection:** preparation that fills gaps or seals them permanently or flexibly
**Intended actions:** activities carried out with the intentional use of a biological agent of known species
**Interferons (IFN):** group of heat-stable soluble basic antiviral glycoprotein cytokines of low molecular weight that are produced by cells exposed to the action of a virus, bacterium or some chemicals
**Interleukins (IL):** various cytokines of low molecular weight that are produced by lymphocytes, macrophages and monocytes and that function especially in the regulation of the immune system and especially cell-mediated immunity
**Intrusion:** forcible entry
**Limulus amoebocyte lysate (LAL):** enzymes extracted from the blood cells of the horse shoe crab (*Limulus polyphemus*) that are activated by endotoxin
**Lipopolysaccharide (LPS):** large molecule consisting of lipids and sugars joined by chemical bonds (see endotoxin)
**Lipolytic:** capable of decomposing lipids and fatty acids and using their decomposition products as a source of carbon and energy
**Lymphocytes:** immune system cells belonging to leukocytes, capable of specific antigen recognition; T lymphocytes help to recognise the antigen; NK lymphocytes have an internal ability to recognise and destroy cells, e.g. infected with virus or cancer
**Lysis:** process of disintegration or dissolution (as of cells)
**Macrophage:** phagocytic tissue cell of the immune system that may be fixed or freely motile, is derived from a monocyte, functions in the destruction of foreign antigens (such as bacteria and viruses) and serves as an antigen-presenting cell
**Mesophilic microorganisms:** microorganisms that can grow in the temperature range of 20–45°C, with optimal temperature between 30–37°C
**Metabolite:** product of metabolism (i.e. chemical changes taking place in organisms)

**Microbiological corrosion:** see bio-corrosion
**Microbiological culture:** method of multiplying microbial organisms by letting them reproduce in a predetermined culture medium under controlled laboratory conditions
**Microbial volatile organic compounds (MVOCs):** chemical compounds of low molecular weight, typically released by growing fungi and bacteria as end-products of their metabolism
**Microbiome:** community of microorganisms (such as bacteria, fungi and viruses) that inhabit a particular environment; also: collective genomes of microorganisms inhabiting a particular environment
**Microbiota:** complex of microorganisms present in a given habitat
**Microorganism:** microbiological entity of any type, cellular or non-cellular, capable of replication or of transferring genetic material, or entities that have lost these properties
**Microwaves:** type of electromagnetic radiation between infrared and ultra-short wavelengths, ranging from 1 mm (frequency: 300 GHz) to 30 cm (1 GHz).
**Mites:** acarid arachnids that often infest animals, plants and stored foods and include important disease vectors, playing a role in allergies caused by house dust
**Monoclonal:** derived from a single clone
**Mould:** microscopic fungus principally producing filaments; grow in the form of hyphae and form a dense mass, called mycelium; asexual spores (conidia) can be easily released into the air
**Muramic acid:** monosaccharide, occurs naturally as N-acetyl derivative of glucosamine in peptidoglycan
**Mutagenicity:** induction of permanent and hereditary changes in the amount or structure of genetic material of a cell or organism
**Mycelium:** vegetative mass of hyphae; thallus body of the fungus
**Mycotoxins:** toxic secondary metabolites produced by fungi
**Nephrotoxicity:** toxic effects on kidneys
**Neutrophils:** granulocyte that is the chief phagocytic white blood cell of the blood
**Occupational exposure:** exposure to potentially harmful chemical, physical, or biological agents that occurs as a result of one's occupation
**Organic Dust Toxic Syndrome (ODTS):** illness occurs as a result of exposure to high levels of microbial agents contained in organic dust
**Peptidoglycan:** polymer that is composed of polysaccharide and peptide chains and is found in bacterial cell walls and some other organisms
**Peroxidases:** group of enzymes catalysing oxidation of hydrogen peroxide of various substrates
**Polyclonal:** product of many different clones of cells
**Polymer:** multiple connections
**Polymerase Chain Reaction (PCR):** key technique in molecular genetics that permits the analysis of any short sequence of DNA (or RNA) without having to clone it; used to reproduce (amplify) selected sections of DNA

**Polysaccharide:** carbohydrate that can be decomposed by hydrolysis into two or more molecules of monosaccharides

**Precipitin:** antibody that forms a precipitate when it unites with its antigens

**Prions:** infectious protein particles devoid of nucleic acids, resulting from mutation processes; can cause chronic human and animal diseases

**Protein:** various naturally occurring extremely complex substances that consist of amino-acid residues joined by peptide bonds, contain elements such as carbon, hydrogen, nitrogen, oxygen, usually sulphur, and occasionally other elements (such as phosphorus or iron), and include many essential biological compounds (such as enzymes, hormones or antibodies)

**Proteolytic:** capable of decomposing proteins, peptides and amino acids and using their decomposition products as carbon and energy sources

**Psychrophilic microorganisms:** microorganisms that can grow in the temperature range 0–20°C, with an optimal temperature of 15–20°C

**Relative humidity (RH):** the ratio of the amount of moisture held in air (vapour pressure) to the maximum amount of moisture that the air can hold at a given temperature and pressure (saturation vapour pressure)

**Remediation:** cleaning and removal of contaminants caused by undesirable effects or failures

**Ribonucleic acids (RNA):** nucleic acid that is used in key metabolic processes for all steps of protein synthesis in all living cells and carries the genetic information of many viruses; unlike double-stranded DNA, RNA consists of a single strand of nucleotides, and it occurs in a variety of lengths and shapes

**Rickettsiae:** mainly rod-shaped, coccoid, and often pleomorphic Gram-negative bacteria that cause febrile diseases in humans and animals, often combined with a rash

**Risk analysis:** the use of available information to estimate the risk to individuals or populations, property or the environment, from hazards

**Risk assessment:** process of making a decision recommending whether or not existing risks are tolerable and present risk control measures are adequate, and if not, whether alternative risk control measures are justified or will be implemented

**Risk:** measure of the probability and severity of an adverse effect on life, health, property or the environment

**Rod:** cylindrical in shape

**Satratoxins:** mycotoxins produced by fungi of *Fusarium* genus

**Sick building syndrome (SBS):** set of adverse symptoms connected with bad quality of indoor air when no specific cause and/or disease could be identified

**Spore:** general term for a reproductive structure in fungi and bacteria

**Teratogenicity:** toxic effects on the embryo or foetus

**Thermal bridge:** part of the building envelope where the uniform thermal resistance is significantly reduced by total or partial penetration of the envelope by materials featuring a different thermal conductivity coefficient, change in thickness of material layers or difference between internal and external surfaces of the partitions

**Thermophilic microorganisms:** microorganisms that can grow in the temperature range 30–90°C, with an optimal temperature of 50–70°C

**Transmissible diseases:** diseases transmitted by invertebrates (most commonly blood-sucking insects and arachnids) referred to as vectors

**Tumour necrosis factor alpha (TNFα):** protein produced by monocytes and macrophages that mediates inflammation and induces the destruction of some tumour cells and the activation of white blood cells

**Unintended actions:** activities where the presence of biological agents, their number, species composition and health risks they may cause are subject to uncertainty

**Vaccine:** preparation of killed microorganisms, living attenuated organisms or living fully virulent organisms that is administered to produce or artificially increase immunity to a particular disease

**Vectors:** in the biological sense, they are invertebrate animals (most often blood-sucking insects and arachnids), transmitting germs of infectious diseases, then referred to as transmissible

**Ventilation:** air exchange or circulation of the air; system or means of providing fresh air

**Viable microorganisms:** microorganisms having a potential for metabolic activity

**Virus:** microorganism that consists of genetic material and a coating and requires living organisms in order to reproduce

**Water activity (aw):** the ratio of the amount of water in a material at particular moisture content (vapour pressure) to the maximum amount of water air can hold at the same temperature and pressure (saturation vapour pressure) (see also: equilibrium relative humidity)

**Xerophilic:** preferring dry places; growing under dry conditions

**Yeast:** fungi, usually single-celled, spherical in shape, with cells that reproduce sexually or asexually by budding; under unfavourable conditions for vegetation, they form spores as dormant structures

**Zoonosis:** disease transmitted from animals to humans

# 1 Water Damage in Buildings and Associated Microbiological Contamination

*Agata Stobnicka-Kupiec*

## CONTENTS



## 1.1 HISTORICAL BACKGROUND OF THE PROBLEM

For centuries, the development of civilisation has been inextricably linked to the processes of microbial colonisation of the places and residences where humans live. The problem of the contamination of indoor environments, and the related phenomenon of biodeterioration of primary products, materials and buildings, has therefore accompanied humanity since the dawn of its history. Probably the oldest mention of the destructive effects of microbiota – referred to back then as red and green 'leprosy' – on rooms and clothing comes from the Book of Leviticus, the third book of the Pentateuch of the Old Testament. Both prehistoric times, as evidenced by analyses of cave paintings from the Paleolithic period, as well as events related to archaeological and conservation research, were associated with properties which were destructive for organic and inorganic materials – mainly mould and actinomycetes, exacerbated by the presence of water in the environment. It was in 23 BC that the famous Roman architect Marcus Vitruvius Pollio, in his work *de Architectura*, made the observation that 'the bedrooms, as well as the libraries, should be situated so as to face eastwards, as this prevents the spread of damp and helps to avoid the spread of mould on both rooms and books'. Among researchers of the modern era, the role of dampness as a factor responsible for indoor bio-corrosion and the associated adverse health effects was probably first highlighted by Iranian doctor

Mohammed bin Zakariya Al-Razi (865–925 AD). He stated that 'avoiding being in damp interiors prevents the sinus pain or rhinitis and reduces airborne infections'. Although the conclusions concerning the importance of water in biodeterioration processes were already formulated in ancient times, significant progress in counteracting the effects of water damage was inhibited until the Renaissance. It was only then that the use of milk of lime as a means of protecting the structural elements of buildings against mould was popularised in Europe by the builders of Italian cities situated by rivers or canals. It was then popularised by Spanish conquistadors in the New World at the end of the 15th century.

Although the biodeterioration of buildings and works of art has been observed since ancient times, and information on the damaging impact of micro-organisms can be found in various periods of history, it has long been believed that the dominant factors of material corrosion are chemical and physical processes, ignoring the importance of biological corrosion [Sterflinger and Piñar 2013]. In practice, only at the end of the 19th century were bacteria and fungi found to have the ability to colonise building materials which may then lose essential properties under their influence. A change in the approach to the problem of bio-corrosion was reflected in the practical measures taken in Europe at that time (mainly in Germany, Austria, Switzerland and France) in the field of so-called 'building mycology'. The work of R. Hartig released in 1885 by the Berlin Springer publishing house, titled *Die Zerstörungen des Bauholzesdurch Pilze. 1. Der ächte Hausschwamm* (*Merulius lacrymans Fr.*), was the inspiration for many initiatives within this field. In Germany and Switzerland, scientific research began on the mechanisms of mould infestation of buildings, as well as the search for appropriate measures to control and prevent it. In 1898, a special committee on the control of household mould was established in Zurich, and seven years later, the German authorities set up a Government Advisory Committee for the Control of Household Mould, consisting of biologists, foresters and construction specialists. As a result of the above, the Institute for Household Mould Research in Wrocław, Poland, was established as a laboratory for scientific research aimed at determining the mechanisms of biological contamination of buildings. Its second director, R. Falck, together with J. Liese and the above-mentioned R. Hartig, created the scientific basis for present-day building mycology. The periods of the First and Second World Wars led to catastrophic destruction and degradation of many buildings, ranging from residential, public and industrial buildings to historic sites and museums. These buildings, overexploited and not renovated during both armed conflicts and the post-war periods, became sites of mass development of microbiological corrosion which led to their biodegradation. At that time, expertise in such degraded buildings was once again an area of interest for researchers, which led to the commercial-scale production of chemicals intended for both the protection of building timber and the removal of mould from walls. The 1930s and the period after the Second World War (1945–1966) led to the development of materials science which, within its field, dealt not only with expert evaluations of biologically corroded buildings, but also with the problems of protection against bio-corrosion. These included the development of methods for the assessment of material protection measures, together with instructions for their defence, and anti-mould protection of buildings [Karyś 2014; Ważny and Karyś 2001].

The processes of microbiological corrosion of materials are inevitable, and the problem of 'sick buildings' and its solution continues to be an important issue for modern science and technology. It requires the interdisciplinary involvement of many fields, combining theoretical and practical knowledge of biology with technical, technological and design methods. These issues are undoubtedly underpinned by a thorough understanding of the phenomenon of bio-corrosion, recognition of its causes and the development of effective ways to prevent it.

## 1.2 CONTEMPORARY PROBLEMS RELATED TO MICROBIOLOGICAL CONTAMINATION OF THE INDOOR ENVIRONMENT

Buildings are continuously exposed to colonisation by micro-organisms present in the environment. During the exploitation of construction objects, their structural elements are subject to environmental stress caused by water in its various forms. Whenever water makes contact with the surface of a construction material, or seeps inside it, microbiological contamination may occur. Roofs, floors, lower walls or foundations are most often exposed to dampness and biodegradation. This type of damage is relatively common and is usually associated with mould, and the scale of this phenomenon is evidenced by numerous scientific works [Górny 2004; WHO 2009].

In recent decades, natural disasters caused by water have hit many regions of the world quite regularly. One of the biggest problems for victims of this type of disaster is the return to their homes and workstations. Buildings damaged by water are usually uninhabitable due to the condition of the structure, damage to various types of installations or sanitary conditions. The presence of water and the pollutants in it create ideal conditions for the development of micro-organisms that can threaten human health, either through direct contact with their source or indirectly through their emission into the air. According to U.S. statistics, almost all of its 119 million homes and 4.7 million public buildings have experienced an episode of severe dampness due to flooding, swamping or water intrusion into their interiors in their history [U.S. Census Bureau 2003]. In 2005, Hurricane Katrina led to a disastrous flood in New Orleans. Destroyed, and abandoned by many inhabitants, the city has become a place where both the ways of spreading microbiological pollutants and the methods of removing them from damp materials can be studied to this day [Adhikari et al. 2009; Chew et al. 2006; Farris et al. 2007; Rao et al. 2007].

Inevitable climate change is causing frequent floods around the world, and thus increasing the risk of water damage to various types of buildings, ranging from residential through public to agricultural and industrial facilities. According to the report published by the European Environment Agency, there were 3,563 floods in 37 European countries between 1980 and 2010. The highest number (321) was recorded in May and June 2010 in 27 Central European countries.

Many regions of the world were devastated by floods in 2019. According to FloodList data, 15,000 houses in India and 13,000 houses in Africa were affected by the floods that occurred in September 2019 alone. The losses incurred by the U.S. as a result of Hurricane Dorian and the floods it caused run into trillions of dollars. At the same time, torrential downpours also led to flooding and inundations in European countries [FloodList 2019]. Scientists predict that by 2050, the number of floods in the world, and the economic losses associated with them, will have increased five times, and by 2080, 17 times. This is due to global warming, increasing the value of land around flood plains and rapid urban development [EEA 2016]. It can therefore be assumed that, considering that number and scale, a long-term effect destroying microbiologically contaminated or damp buildings and potentially causing health problems may affect families whose dwellings did not undergo appropriate restoration and drying procedures, as well as appropriate anti-mould protection. The scale of this problem is also evidenced by the fact that the cost of damage to buildings caused by microbiological corrosion is estimated to exceed €200 million per year in Germany alone [Sedlbauer 2001]. According to Finnish data, the estimated unit cost of repairing microbiologically damaged interiors which had an adverse impact on the health of occupants may be as high as €10,000–€40,000 per year [Pirinen et al. 2005]. Nevertheless, the overall costs of removing the bio-corrosion from building construction are difficult to estimate. This is because they include not only the costs of cleaning, but also renovation and repair, and, in many cases, also the costs of destruction of cultural assets associated with the devastation of historic buildings [Gaylarde et al. 2003].

Undoubtedly, it can be stated that the problem of microbiological contamination of interiors is nowadays a worldwide phenomenon, and building materials are affected by biodeterioration regardless of the climate zone or location of the building [Ribas Silva 1996]. According to a survey carried out in North America, 27–36% of houses are affected by mould. According to research using measurements of indoor air quality, this proportion reaches even 42–56%. In European countries, the proportion of damp and mouldy residential buildings varies between 12–46% in the UK, 15–20% in the Netherlands and Belgium and 12–32% in the Nordic countries (Sweden, Denmark, Finland, Norway, Iceland and Estonia). Despite this, the proportion of signs of building damage caused by excessive dampness, as recognised by construction engineering specialists, can reach up to 80% of residential buildings. A similar situation is observed in the Middle East and in Asian countries. For example, in Israel as many as 45% of houses have problems with dampness and mould, while in 20% of the interiors these problems are considered serious. In the Gaza Strip and on the West Bank of the Jordan River, it was observed that in 56% of the houses the mould spread on walls and ceilings. In Ramallah it was found that this proportion is as high as 78%. It is also estimated that in agricultural areas of Taiwan 12% of homes are damp, 30% have signs of household mould, 43% have had water intrusion and in 60% of interiors at least one of the above-mentioned signs occurred. In Japan, almost 16% of houses show signs of mould [Adhikari et al. 2009; Chew et al. 2006; Farris et al. 2007; Rao et al. 2007].

The consequences of microbiological contamination of premises are, on the one hand, progressive corrosion of building materials and, on the other hand, a decrease in the quality of indoor air, resulting in the occurrence of 'sick building syndrome' (SBS) symptoms such as fatigue, headaches, dizziness, nausea, exhaustion or mucous membrane irritation [Sundell et al. 1994; Robertson 1988].

### 1.2.1 Traditional and Modern Building Materials

The comfort of the building and its resistance to biodeterioration depends largely on the building materials, what they are made of and their proper use. Despite significant progress in the construction sector, there is no clear methodology for implementing an appropriate selection of these materials [Fezzioui et al. 2014]. Among building materials we can distinguish traditional and modern ones. The concept of traditional materials is twofold: they are materials that have been used for centuries for construction purposes and that are referred to as natural or ecological (e.g. wood, stone, burnt brick, reeds or straw). Modern materials are materials that have only been known in construction for several decades (cement, brick, breeze-blocks, concrete) (Figure 1.1). Nowadays, in modern buildings the importance of decorative and ecological aspects of traditional building materials is increasing [Wang 2014]. Ecology, energy-saving, low costs and the possibility of using local raw materials make wood, stone, straw or bamboo very popular in construction.

It should be noted that bacteria and fungi have the ability to colonise a variety of building materials and cause their biodeterioration which, in turn, is increased by acid corrosion, enzymatic degradation and mechanical stress [Sterflinger and Piñar 2013]. These phenomena may concern various materials including stone, concrete, mortars, suspensions and paint coatings, as well as glass and metals, that are used in many architectural solutions [Piñar and Sterflinger 2009]. Organic materials such as wood, reed or bamboo, due to their cellulose content, are rich sources of nutrients for micro-organisms (mainly fungi) producing ligninolytic and cellulolytic enzymes, and in this respect are more likely to undergo biodeterioration. On the other hand, bacteria also play a major role in the initiation of biodeterioration of metals which are an example of modern building materials, and the process of their degradation is

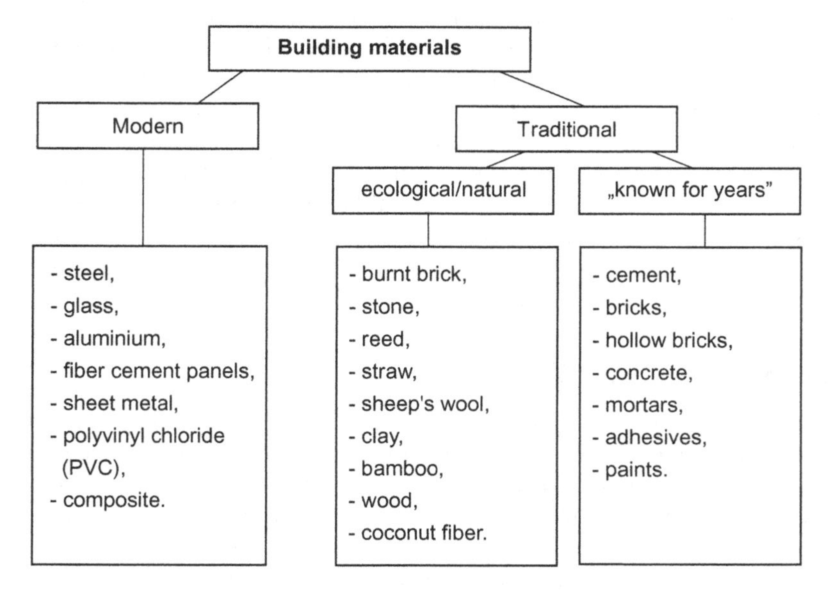


**FIGURE 1.1** Types of building materials used in modern construction.

primarily associated with acid corrosion [Gaylarde and Morton 1999]. Materials with porous structures, such as brick, concrete or mortar, are also colonised by micro-organisms, and the mineral salts in them can be dissolved by microbial metabolites or used directly by their cells as nutrients [Gaylarde et al. 2003].

### 1.2.2 Aspects of Energy Economy

Optimising the energy efficiency of the building, while ensuring the comfort and health of the occupants, is closely linked to the microbiome of the interior. Quantitative and qualitative 'management' of micro-organisms' indoor environment can reduce the biodegradation of building materials and thus, for example, reduce the formation of biofilm on them, which minimises energy losses. However, achieving these goals is not easy. For example, raising the temperature in water heaters and pipes installed in a building to a level that inhibits microbial growth can result in higher energy costs and, at the same time, has limited effectiveness when it comes to inactivating certain micro-organisms. On the other hand, an increase in the inflow of external air delivered into the building is associated with greater comfort of its occupants but, at the same time, raises exposure to a greater variety of micro-organisms and increases energy consumption for heating and cooling of that building [NAS 2017].

Modern construction places emphasis on environmental care through aspects of energy economy. In order for a building to meet the energy efficiency requirement, it must have, among other things, adequate thermal insulation. In a building, heat can escape through walls, roof and floor. However, the most important place where heat escapes from buildings are windows and doors as well as their connection points to the walls [BPIE 2010]. Incorrect installation of windows often leads to the formation of the so-called 'thermal bridge' – a part of the building which has much worse thermal insulation than the adjacent part. Thermal bridges have adverse effects, as they lead to heat loss and point or linear cooling of parts of the building envelope. This may be conducive to condensation and dampness, which creates favorable conditions for microbial growth [Gorse and Johnston 2012]. Other consequences of thermal bridges include uncontrolled heat loss, up to 30%, and mechanical damage to the structure. Structural nodes that connect different elements of the building envelope are mostly prone to the occurrence of thermal bridges, and thus microbial growth. Among such places are the connection points between the roof and external wall, as well as between the balcony and ceiling, window frames, ring beams, lintels and cellar walls [Ge et al. 2013; Allen and Lano 2009].

### 1.2.3 Modern 'Green' Building

'Green' building, also referred to as ecological or sustainable building, is an activity related to the design and construction of buildings in accordance with the principle of 'sustainable development', particularly manifesting in care for the environment, as well as in the economic management of raw materials throughout the entire construction cycle, starting from the design of the building, through construction work and use (including proper maintenance and modernisation) to its demolition [Fischer 2010]. The concept of setting building standards and rules, which are based on the elements that are least harmful to the environment during the whole period of the buildings' operation, consists, among other things, of the use of environmentally friendly building materials and reduction of energy consumption, as well as greening the buildings and the areas adjacent to them [Mikoś 2006]. Building materials used in accordance with the philosophy of 'green' building must be 'environmentally friendly', and therefore recyclable and derived from materials of biological origin (bio-based products). Sustainable building is therefore a form of construction that is designed to take into account the use of easily accessible, reusable or recycled materials [Muazu and Alibaba 2017]. Thus, the philosophy of 'green' building assumes the possibility of using not only traditional materials for construction, but also materials derived from waste obtained from scrapyards and landfills, as well as building elements intended for demolition. Wood, plasterboards or stone tiles are often reclaimed building materials. The house of American entrepreneur Jacek Helenowski is an example of a 'green house'; it was made almost entirely of reclaimed or recycled materials and, as one of the greenest homes in the U.S., was given an award by the U.S. Green Building Council [Green Home Institute 2010]. In such cases, however, it should always be taken into account that unwanted micro-organisms may be present on reclaimed building materials, and thus such structural elements may become a source of microbiological contamination of building interiors.

Also, some construction solutions applied in 'green' building can contribute to microbiological contamination of the indoor environment. A common construction solution in ecological buildings is the use of green roofs (inverted roofs), which are special roof coverings consisting of multiple layers, including waterproofing, thermal insulation, protective drainage and filtering layers, with earth on top on which plants may be grown. This type of roofing has a number of advantages: among others, it muffles all noise very well, prevents major heat losses in winter and protects the building against excessive heating in summer, as well as helping to reduce the occurrence of the so-called 'urban heat island' phenomenon. Nevertheless, this solution also has some disadvantages. Above all, design and construction errors can contribute to the leakage of water into the building and the penetration of plant roots through the roofing, which can lead to microbiological corrosion of the building [Fischer 2010]. The studies conducted so far have shown the presence of both bacteria and fungi in green roof ecosystems, among which there are numerous species decomposing organic matter, which, in the case of faulty roof construction and its leakage, may threaten the remaining building envelopes [McGuire et al. 2015].

Although 'green' building has the potential to improve indoor air quality and, thus, have beneficial impact on the health of occupants, no studies have so far been carried out to determine how a change of residence from a conventional to a 'green' building is linked to health effects. Therefore, before science approaches this matter with carefully proven data, it should be borne in mind that many of the characteristics of 'green' building may not fully contribute to improving the quality of the indoor environment and, thus, the health of occupants [NAS 2017].

## REFERENCES

Adhikari, A., J. Jung, T. Reponen et al. 2009. Aerosolization of fungi, (1–3)-β-D glucan, and endotoxin from flood-affected materials collected in New Orleans homes. *Environmental Research* 109(3):215–224.

Allen, E., and J. Lano. 2009. *Fundamentals of Building Construction: Materials and Methods*. Hoboken, NJ: John Wiley & Sons.

BPIE [Buildings Performance Institute Europe]. 2010. *Cost Optimality: Discussing Methodology and Challenges Within the Recast Energy Performance of Buildings Directive*. https://www.buildup.eu/sites/default/files/content/BPIE_costoptimality_publication2010.pdf. [accessed October 11, 2019].

Chew, G. L., J. Wilson, F. A. Rabito et al. 2006. Mold and endotoxin levels in the aftermath of Hurricane Katrina: A pilot project of homes in New Orleans undergoing renovation. *Environmental Health Perspectives* 114(12):1883–1889.

EEA [European Environment Agency]. 2016. *Floodplain Management: Reducing Flood Risks and Restoring Healthy Ecosystems*. https://www.eea.europa.eu/highlights/floodplain-management-reducing-flood-risks. [accessed October 11, 2019].

Farris, G. S., G. J. Smith, M. P. Crane, C. R. Demas, L. L. Robbins, and D. L. Lavoie. 2007. *Science and the Storms: The USGS Response to the Hurricanes of 2005*. Reston, VA: U.S. Geological Survey.

Fezzioui, F., M. Benyamine, S. Larbi, B. Draoui, and C. A. Roulet. 2014. Impact of traditional and modern building materials on the thermal behavior and energy consumption of a courtyard house in the hot and dry climate. *International Conference On Construction Materials And Structures (ICCMATS)*, 24–26 November 2014, Johannesburg.

Fischer, E. A. 2010. Issues in green building and the federal response: An introduction. https://www.researchgate.net/publication/326059820_Issues_in_green_building_and_the_federal_response_An_introduction. [accessed October 11, 2019].

FloodList. http://floodlist.com. [accessed October 11, 2019].

Gaylarde, C. C., and L. H. G. Morton. 1999. Deteriogenic biofilms on buildings and their control: A review. *Biofouling* 14(1):59–74.

Gaylarde, C., M. Ribas Silva, and T. Warscheid. 2003. Microbial impact on building materials: An overview. *Materials and Structures* 36(5):342–352.

Ge, H., V. R. McClung, and S. Zhang. 2013. Impact of balcony thermal bridges on the overall thermal performance of multi-unit residential buildings: A case study. *Energy and Buildings* 60:163–173.

Górny, R. L. 2004. *Cząstki grzybów i bakterii jako składniki aerozolu pomieszczeń: Właściwości, mechanizmy emisji, detekcja*. Sosnowiec: Wyd. IMPiZŚ.

Gorse, C. A., and D. Johnston. 2012. Thermal bridge. In *Oxford Dictionary of Construction, Surveying, and Civil Engineering*, eds. Gorse, C., D. Johnston, and M. Pritchard. 3rd ed, 440–441. Oxford: Oxford UP.

Green Home Institute. 2010. https://greenhomeinstitute.org/new-leed-certification-leed-platinum-net-zero-in-chicago. [accessed October 11, 2019].

Hartig R. 1885. *Die Zerstörungen des Bauholzesdurch Pilze. 1. Der ächte Hausschwamm (Merulius lacrymans Fr.)*.

Karyś, J., ed. 2014. *Ochrona przed wilgocią i korozja biologiczna w budownictwie*. Warsaw: Grupa Medium.

McGuire, K., S. Payne, G. Orazi, and M. Palmer. 2015. Bacteria and fungi in green roof ecosystems. In *Green Roof Ecosystems: Ecological Studies: (Analysis and Synthesis)*, ed. R. Sutton, Vol. 223. Cham: Springer.

Mikoś, J. 2006. *Budownictwo ekologiczne*. Gliwice: Wyd. Politechniki Śląskiej.

Muazu, A. G., and H. Z. Alibaba. 2017. The use of traditional building materials in modern methods of construction: (A case study of Northern Nigeria). *The International Journal of Monitoring and Surveillance Technologies Research* 2(6):30–40.

NAS. [National Academies of Sciences, Engineering, and Medicine]. 2017. *Microbiomes of the Built Environment: A Research Agenda for Indoor Microbiology, Human Health, and Buildings*. Washington, DC: The National Academies Press.

Piñar, G., and K. Sterflinger. 2009. Microbes and building materials. In *Building Materials: Properties, Performance and Applications*, eds. D. N. Cornejo, and J. L. Haro, 163–188. New York: Nova Science Publishers.

Pirinen, J., J. Karjalainen, J. P. Kärki, H. Öhman, and T. Riippa. 2005. Homevauriot suomalaisissa pientaloissa. *Espoo, Sisailmastoseminaari.* (SIY Report 23).

Rao, C. Y., M. A. Riggs, G. L. Chew et al. 2007. Characterization of airborne molds, endotoxins, and glucans in homes in New Orleans after Hurricanes Katrina and Rita. *Applied and Environmental Microbiology* 73(5):1630–1634.

Ribas Silva, M. 1996. Climates and biodeterioration of concrete. In: *Durability of Building Materials & components 7*, ed. C. Sjöström, T. 1, 191–200. London: Taylor & Francis.

Robertson, G. 1988. Source, nature and symptomology of indoor air pollutants. In *Indoor and Ambient Air Quality*, eds. R. Perry, and P. W. Kirk. London: Publications Division Selper.

Sedlbauer, K. 2001. *Prediction of Mould Fungus Formation on the Surface of and inside Building Components.* Stuttgart: Fraunhofer Institute for Building Physics.

Sterflinger, K., and G. Piñar. 2013. Microbial deterioration of cultural heritage and works of art-tilting at windmills? *Applied Microbiology and Biotechnology* 97(22):9637–9646.

Sundell, J., T. Lindval, and S. Berndt. 1994. Association between type of ventilation and airflow rates in office buildings and the risk of SBS-symptoms among occupants. *Environment International* 20(2):239–251.

USCB [United States Census Bureau]. 2003. https://www.census.gov. [accessed, October 11, 2019].

Vitruvius Pollio, M. c. 80/70 – c. 25 B.C. *De Architectura*, Available at: http://www.thelatinlibrary.com/vitruvius.html

Wang, J. 2014. The application of traditional building materials in modern architecture. *Applied Mechanics and Materials* 644–650:5085–5088.

Ważny, J., and J. Karyś. 2001. Ochrona budynków przed korozją biologiczną. *Arkady, Warszawa* 13–19:52–90.

WHO [World Health Organization]. 2009. *Guidelines for Indoor Air Quality: Dampness and Mould.* Copenhagen: World Health Organization Regional Office for Europe.

# 2 Indoor Microbial Pollutants

*Marcin Cyprowski, Anna Ławniczek-Wałczyk, Rafał L. Górny, and Agata Stobnicka-Kupiec*

**CONTENTS**



## 2.1 FUNGI

Fungi are ubiquitous in the environment. They occur primarily in the soil, on decaying or dead organic matter. They are able to produce a large number of conidia, release them into the air and, by that, easily colonise new areas. The amount of fungal conidia exceeds the amount of pollen and bacterial spores in the air [Adan 1994]. This dominance is ensured by the enormous productivity of the mycelium, the ease of conidia release and their ability to survive (even several decades of) a drying period [Davis 2001; Mandrioli et al. 2003]. As a result of natural (e.g. through the open windows, doors etc.) and forced (ventilation) atmospheric air infiltration, fungal conidia are constantly present in buildings. The qualitative composition of indoor microbiota usually reflects the species composition of outdoor air [Miller 1992]. Quantitatively, the period from spring to autumn is characterised by a lower load of fungal conidia in the indoor air compared to the winter period, when their number exceeds the level recorded in the outdoor air [Reponen et al. 1992]. This type of situation is usually encountered in so-called healthy rooms, i.e. those in which residents do not complain about ailments which in their opinion would be caused by the sanitary condition of premises they live in and in which no additional sources of bioaerosols (e.g. mouldy surfaces) are present.

In buildings we also often deal with spatial and temporal variations in the concentration of microbiological agents. In the case of fungi, spatial differences occur when a significant internal source of their emission is present, situated in a specific part(s) of the building and when the air exchange between the premises within the building does not ensure the maintenance of homogeneous conditions. In turn, time changes in fungal concentrations depend on: air exchange rate, time of day and activity of people occupying certain areas of the building. Such considerations should also take into account the above-mentioned seasonality of the fungal life-cycle and the biomicroclimatic parameters of the premises. Broader discussions regarding spatial and temporal qualitative and quantitative changes of indoor mycobiota were already published by Lehtonen et al. [1993], Lighthart and Stetzenbach [1994], Hyvärinen et al. [2002] or Herbarth et al. [2003].

There are about 100,000 known fungal species, which, due to their morphology and having specific structures that provide an appropriate way of reproduction, are classified into a total of four to seven (depending on the classification used) types (*Phyla*) (e.g. de Hoog and Guarro 1995; St-Germain and Summerbell 1996; Fisher and Cook 1998; Davis 2001; Kirk et al. 2011; Watkinson et al. 2015). With a few exceptions, among the fungi that have the closest connection with the indoor environment, the majority are *Deuteromycota* type fungi [Miller 1992]. In the scientific literature, there are numerous works characterising the building microbiota. Their analysis revealed that among moulds, species of *Penicillium*, *Aspergillus*, *Cladosporium* and *Alternaria* genera are the most common indoors, both in the air and on surfaces (e.g. Hunter et al. 1988; Miller et al. 1988; Grant et al. 1989; van Reenen-Hoekstra et al. 1991; Hyvärinen et al. 1993; Beguin and Nolard 1994; Beguin 1995; Gravesen et al. 1999; Morey 1999; Hiipakka and Buffington 2000; Flannigan and Miller 2001; Górny and Dutkiewicz 2002; Gots et al. 2003; Kemp et al. 2003). In buildings where water damage has occurred, species from *Stachybotrys*, *Chaetomium*, *Fusarium*, *Trichoderma* and *Paecilomyces* genera are also relatively common [Pasanen 1992a, 1992b].

### 2.1.1 Fungal Allergens

Fungal allergens are the main cause of atopic diseases [Kurup and Banerjee 2000]. According to various authors, from 80 to over 100 species of fungi are linked causally to the symptoms associated with allergic respiratory diseases [Horner et al. 1995; Helbling and Reimers 2003]. Fungal allergens are mostly proteins with molecular weight from 10,000 to 80,000 daltons (Da) [Larsen 1994; Kurup and Banerjee 2000], although most of the allergen extracts found today are a mixture of proteins, glycoproteins, polysaccharides and other substances [Esch 2004].

Each fungal species can produce dozens of allergens. Studies show that the allergen content of a particular fungal species depends on the age of its colony, including the number of culture transfers of a particular microorganism, the temperature, the substrate on which it grows, and even the strain within that species [Larsen 1994; Bisht et al. 2000]. The survival of conidia also influences the number of released allergens. In the case of *Aspergillus* allergens, their amount increases significantly during the gemination process [Sporik et al. 1993]. In *Alternaria*, variability in allergen content between individual conidia, between individual strains of the same fungal species and between conidia and hyphae isolates of the same species has been documented [Portnoy et al. 1993]. The *Asp* f1 allergen is released or secreted from germinating conidium more numerous than from the non-germinating spore, as has already been shown for the *Aspergillus* allergen [Sporik et al. 1993]. The growing tip of the conidium contains protein allergens, which was demonstrated by gold labeling in the IgE reaction against *Aspergillus fumigatus* allergens [Reijula et al. 1991] and observed through the intense staining of the *Alternaria* allergen at the ends of hyphae of germinating conidia [Matakakis et al. 2001].

While the allergens released from the mixture of conidia and hyphae fragments are recognised, as is the case with *Alternaria* allergens, their identification during emission in extracts containing separated conidia and mycelium has so far been insufficient. In a few studies carried out for this purpose, allergic responses caused by pure (96–97%) extracts of both these structural elements were compared. However, these studies are not unambiguous in their meaning. Their results allowed identification in both extracts not only of antigens common to both structural elements of the fungus, but also antigens of different reactivity, characteristic only for spores and mycelium [Aukrust et al. 1985; Fadel et al. 1992].

While some studies have shown that conidia contains specific allergens that are more reactive in skin tests than those derived from the mycelium [Solomon et al. 1980; Hoffman et al. 1981] or the reactivity of conidia allergens is greater than that of mycelium due to their loss in this area [Licorish et al. 1985], in other cases, in patients the intensity of the reaction to the mycelium extracts exceeded that of the conidia extracts [Aukrust et al. 1985; Fadel et al. 1992], probably due to the fact that the main *Alternaria* allergen was present in more quantities in mycelium than in conidia [Paris 1990b]. Moreover, studies on mycelium and conidium extracts have shown that the former may contain 40% more secondary metabolites (including mycotoxins) [Fisher et al. 2000].

### 2.1.2 (1→3)-β-D-Glucans

Glucans are water-insoluble components of the cell wall of most fungi and some bacteria. In terms of chemical structure, glucans are glucose polymers which are classified as α or β according to the type of intra-chain bonding. In the fungal cell wall there are densely packed glucan networks connected by (1→3)- and (1→6)-β-D-glycosidic bonds, which with such proteins, lipids and carbohydrates as mannan or chitin provide the cell wall with rigidity and integrity [Douwes 2005; Rylander et al. 1992; Ławniczek-Wałczyk and Górny 2010]. The first scientific reports indicating (1→3)-β-D-glucans as a factor responsible for the adverse health effects associated with an indoor environment contaminated with mould appeared in the late 1990s [Douwes 2005; Rylander et al. 1992]. Attention was also drawn to the ability of (1→3)-β-D-glucans to modify the body's immune response to endotoxins and other environmental allergens, which may lead to an increase in the incidence of adverse health effects in people living in microbiologically contaminated premises [Douwes 2005; Ławniczek-Wałczyk and Górny 2010]. It is also believed that the harmful properties of glucans do not depend on the viability of fungi, and those released from dead organisms or their fragments may have the same negative impact on human health [Douwes 2005; Ławniczek-Wałczyk and Górny 2010].

Workers employed in the renovation of flood-affected houses are particularly exposed to harmful moulds and β-glucans. During demolition, drying and disinfection of water-destroyed houses, β-glucans are often aerosolised from mould-contaminated surfaces. Studies on the phenomenon of harmful microbiological agent aerosolisation from the surface of various finishing materials collected from flood-affected houses (fragments of linoleum, carpets, mattresses and pillows) showed that the concentrations of β-glucans released into the air from such materials are at the level of 2–29 $\mu g/m^3$ [Adhikari et al. 2009]. In the study by Andersson et al. [1997], the concentration of β-glucans in a sample of mouldy gypsum board was in the range of 2.1–14 × $10^3$ μg/g, and in mineral wool samples their concentration was 2.5 μg/g on average. Other researchers noted that the average levels of β-glucans in houses where renovation works are carried out reached 118 $\mu g/m^3$ and were almost four times higher than in houses where renovation has long been completed [Hoppe et al. 2012; Rando et al. 2014].

### 2.1.3 Mycotoxins

Mycotoxins are a chemically heterogeneous group of low molecular weight substances (200–400 Da), produced by certain moulds (including those of the genera *Aspergillus*, *Penicillium* and *Fusarium*) and characterised by their toxicity to animals and humans. These substances can be spread in the air by means of conidia, fragments of mycelium or a medium on which fungi grow [Fisher et al. 2000]. They are often produced during fungal growth and released into the environment in large quantities when the colony lacks nutrients and water. Due to their chemical structure and the resulting specific biological properties, mycotoxins can be divided into several groups [Soroka et al. 2008], of which the greatest risk is posed by:

a) aflatoxins – produced by *Aspergillus flavus* and *A. parasiticus*; they exhibit carcinogenic, mutagenic and teratogenic effects;
b) ochratoxins – produced by *Aspergillus ochraceus* and *Penicillium verrucosum*; they have mainly nephrotoxic effects;
c) zearalenone produced by species of the genus *Fusarium* and trichothecenes produced by species of the genera *Fusarium*, *Cephalosporium*, *Myrothecium*, *Trichoderma* and *Stachybotrys*; they have immunomodulatory and immunotoxic effects, and are, inter alia, inhibitors of protein synthesis.

The difficulties in assessing exposure to mycotoxins in buildings that have experienced water damage are due to two main reasons. Firstly, not all strains of fungi that have the potential to produce mycotoxins actually produce them. As shown by Bloom et al. [2009], who examined dust samples from the floor surfaces of houses affected by Hurricane Katrina, the concentrations of mycotoxins did not correlate with the concentrations of the determined moulds. It is assumed that this may be due to interactions between different fungi species as well as other microorganisms. According to Chełkowski [1985], co-occurrence of *Aspergillus flavus* with *A. niger*, *A. chevalieri*, *A. candidus* and *Trichoderma viride* may completely inhibit its ability to produce aflatoxins. In addition, the production of mycotoxins may be affected by humidity, temperature and even geographical location. Methodological considerations are another important reason for the observed discrepancies. There is no single, generally accepted method of assessing mycotoxin concentrations. Due to the expected high concentrations, settled dust samples are often taken; however, such samples may differ in quality (different microbial spectrum) from those taken from the air. The analytical methods used may also limit the identification of mycotoxins to some extent. Gas or liquid chromatography in combination with mass spectrometry allow the exact determination of the types of mycotoxins and their concentrations; however, these are expensive procedures and therefore rarely used. More practical, although slightly less accurate in quantitative terms, are the immunoenzymatic methods based on the specific reaction between antigen and antibody [Brasel et al. 2005; Charpin-Kadouch et al. 2006].

Of the numerous moulds inhabiting building materials, significant mycotoxin production has been shown for only a few species, including *Stachybotrys chartarum* and *Aspergillus versicolor.* According to the available data, *Penicillium* moulds do not produce large quantities of these compounds [Nielsen 2003]. On the basis of environmental studies, it can be concluded that in buildings the concentration of trichothecenes and sterigmatocystin on wood, wallpaper and cardboard, associated with the colonisation of the studied areas by *S. chartarum*, varies between 1–15 $\mu g/cm^2$, and associated with the contamination by *A. versicolor* between 1–23 $\mu g/cm^2$. In a study by Gottschalk et al. [2008], the concentration of satratoxin-H on the wallpaper surface reached 12 $\mu g/cm^2$, and in air samples from an apartment where *S. chartarum* was previously found, the concentration of satratoxins-G and -H were 0.25 $ng/m^3$ and 0.43 $ng/m^3$, respectively. Assessments of mycotoxins in indoor air have rarely been conducted so far. In the Brasel et al. [2005] study, the air samples were tested in seven fungal-contaminated buildings, where the concentrations of trichothecenes determined using ELISA ranged between 10–1300 $pg/m^3$.

### 2.1.4 Microbial Volatile Organic Compounds (MVOCs)

Microbial volatile organic compounds (MVOCs) are chemical compounds (usually aldehydes, alcohols, ketones, terpenes, esters, amines) of low molecular weight which are released into the air as a result of metabolic reactions of fungi and bacteria present in this environment [Wilkins et al. 2000]. They are produced during intensive microbial growth and are characterised by low concentrations. The researchers consider them primarily as chemical indicators of mould growth in indoor areas. Research indicates that the production of specific MVOCs is highly dependent on the fungal species which can be used as a means to identify indoor mycobiota [Kuske et al. 2005].

It is estimated that so far, more than 200 different MVOCs have been identified in laboratory tests, a significant proportion of which have been determined in building materials inhabited by moulds of *Absidia, Acremonium, Alternaria, Aspergillus, Botrytis, Chaetomium, Cladosporium, Coniophora, Fusarium, Paecilomyces, Penicillium, Phialophora, Rhizopus, Stachybotrys, Trichoderma, Ulocladium* and *Wallemia* genera [Korpi et al. 2009]. However, in humid or mouldy buildings only about 15 different MVOCs are most often analysed [Ström et al. 1994], among which there are compounds with a characteristic, perceptible scent:

- geosmin – an earthy scent, perceptible at a concentration of 150–200 ng/m$^3$;
- 1-Octene-3-ol – scent of fresh mushrooms, perceptible at a concentration of ~10 μg/m$^3$;
- 2-Octene-1-ol – a musty/rotten scent, perceptible at a concentration of ~16 μg/m$^3$.

The possibility of perceiving MVOCs by smell is different for each person. Nevertheless, in environmental studies the application of such ways of recognising compounds are being tried. Keller et al. [1998] estimate that it is difficult to recognise a 'fungal-like' smell below 0.035 μg/m$^3$. At concentrations between 0.05–1.72 μg/m$^3$, it is usually recognisable as a mild fungal scent, and a strong fungal scent (odour) can be associated with concentrations reaching ~12.3 μg/m$^3$. However, it should be remembered that these compounds are also naturally present in the outdoor environment, where their concentrations may vary between 1.1–9.5 μg/m$^3$ [Ström et al. 1994] and from where they may migrate into the premises. Testing of MVOCs levels in buildings is not frequent as there is still no standardized method for their analysis. They are most often examined using the gas chromatography technique combined with mass spectrometry [Keller et al. 1998].

## 2.2 BACTERIA

### 2.2.1 Actinomycetes and Their Allergens

Aerobic actinomycetes are a group of filamentous bacteria that have developed an unprecedented variety of forms and functions compared to other microorganisms. The number of *Streptomyces* species alone is estimated at over 31,000 [Zaremba and Borowski 2001]. These microorganisms had a unique ability to colonise the so-called 'hard' surfaces. Their ability to survive on rocks, plants, animals, clothing, foodstuffs and other uncovered surfaces is due to the property of the spores to survive long drying periods with low moisture content in the substrate on which they grow [Ensign 1978; Williams et al. 1989].

Although much less frequently examined by aerobiologists than fungi, actinomycetes are relatively common in indoor air [Lacey 1988]. Studies carried out in the United States and Europe have shown that their prevalence in indoor environments which do not show microbiological contamination ranges from 2% to 19% and their concentration does not exceed 30 CFU/m$^3$ [Hirsch and Sosman 1976; Górny 1998]. Actinomycetes are also detected in premises where the presence of odours from corroded building materials is found and their levels in the air of damp flats can reach 70% [Nevalainen et al. 1991; Sunesson et al. 1997].

A particularly interesting group among the actinomycetes is *Streptomyces* genus. These Gram-positive spore-forming bacteria are characterised by high resistance to stress caused by dehydration and unusual metabolic activity [McBride and Ensign 1987]. They are capable of synthesising more than half of the 10,000 documented bioactive compounds [Anderson and Wellington 2001]. Mesophilic *Streptomycetes* were isolated from buildings damaged by moisture, where they grew on the surfaces of construction and finishing materials (especially ceramic materials and gypsum boards) [Hyvärinen et al. 2002]. Since they do not belong to the normal microbiota of this type of indoor environment, their presence is considered to be an indicator of the areas polluted due to water damage [Hirsch and Sosman 1976; Dutkiewicz et al. 1988; Cole et al. 1994].

Actinomycete allergens are not as well-known as fungal allergens. So far, only a few species have been more closely characterised. These include mainly allergens of thermophilic species such as *Saccharopolyspora rectivirgula* (syn. *Micropolyspora rectivirgula, Faenia rectivirgula*) or *Thermoactinomyces sacchari. S. rectivirgula* allergens reach a molecular weight of $39 \times 10^3$ Da to $265 \times 10^3$ Da and contain both protein and sugar components. Since certain antigens are very sensitive to temperature, it is suggested that they are destroyed e.g. during the hay and precipitin heating process. They are found, for example, in the serum of patients with a farmer's lung disease and are probably a result of the spore gemination taking place in the lungs [Lacey 1981].

### 2.2.2 Bacteria as a Source of Peptidoglycans

The importance of bacteria, including Gram-positive bacteria (from the cocci, corynebacteria or bacilli group), as factors determining human well-being in the environment has not been fully understood so far, although these bacteria clearly dominate in the indoor microbiota [Gołofit-Szymczak and Górny 2018]. It should be emphasised that their potential health harmfulness cannot be assessed solely on the basis of infectious or allergic properties. An equally important and still underestimated route of exposure is inhalation of immunologically active airborne peptidoglycans. Peptidoglycans are an important structural component of the bacterial cell wall, consisting of a muramic acid biopolymer and N-acetylglucosamine, connected by $\beta$-(1$\rightarrow$4)-glycosidic and pentapeptide bonds. It is estimated that in Gram-positive bacteria, peptidoglycans constitute about 70% of the whole cell wall, whereas the respective percentage in Gram-negative bacteria is about 25% [Sigsgaard et al. 2005].

The use of peptidoglycans as an exposure marker has so far been limited to individual studies in pig farms [Jolie et al. 1998], waste sorting plants [Laitinen et al. 2001; Cyprowski et al. 2019], sewage treatment plants [Cyprowski et al. 2019], metal plants [Cyprowski et al. 2016], in biomass processing [Sebastian et al. 2006], in plant production [Góra et al. 2009] and in office buildings [Cyprowski et al. 2019]. Knowledge about adverse effects of peptidoglycans on the human body is therefore still limited. According to the researchers, they can, similarly to endotoxins, induce the production of such proinflammatory mediators as IL-1, IL-6, IL-8, TNF$\alpha$ or IL-12 [Myhre et al. 2006].

### 2.2.3 Gram-Negative Bacteria as a Source of Endotoxins

In buildings affected by dampness or flooding, Gram-negative bacteria grow particularly well [Andersson et al. 1997]. One of the most characteristic structures of their outer cell wall, selectively recognised by the cells of human non-specific immune response, is the endotoxin. Chemically, it is a macromolecular lipopolysaccharide (LPS), which is released into the environment through fragmentation of the cell wall. A single LPS molecule is composed of three separate regions: O chain, oligosaccharide core and lipid A, which is the centre of endotoxin biological activity [Ławniczek-Wałczyk and Górny 2010]. The endotoxin molecules inhaled together with the dust activate non-specific pulmonary macrophages, which secrete numerous substances with strong biological effects, referred to as proinflammatory mediators (e.g. IL-1, IL-6, IL-8 and TNFα).

Studies show that endotoxin concentrations in flooded buildings vary widely from 0.6 to 139 endotoxin units (EU) in 1 $m^3$ of the air [Solomon et al. 2006; Chew et al. 2006; Rao et al. 2007] and from $7 \times 10^2$ to $9.3 \times 10^4$ EU per 1 $m^2$ of flood-contaminated material (e.g. linoleum, carpets, rugs, mattresses and pillows) [Adhikari et al. 2009]. Endotoxin concentrations in microbiologically corroded gypsum board and mineral wool samples may be at levels of $8.5 \times 10^3$ EU/g and 50 EU/g, respectively. It should be emphasised that the adverse effects of endotoxins on the body persist even after the death of bacterial cells. The LPS secreted in this way from the cell wall is still biologically active. Hence, bacterial endotoxins are one of the important and objective indicators of environmental contamination with Gram-negative bacteria [Liebers et al. 2008; Ławniczek-Wałczyk and Górny 2010].

## 2.3 VIRUSES

Viruses are the smallest biological agents that can lead to indoor contamination. These are simple structures that do not have a cellular structure and do not exhibit features of living organisms – they are not divided outside the host cell and they do not synthesise proteins themselves, nor do they replicate their genome. It is not until the virus enters the body that the consequences of this process in the form of its multiplication and pathological changes lead to the development of the disease. The single active particle of the virus, the so-called 'virion', consists of nucleic acid (DNA or RNA) surrounded by a protein shell (capside) and, if present, a phospholipid membrane which is a part of the host cell. Virus particles vary in size and range between 10–400 nm [Heczko et al. 2014; Stobnicka-Kupiec and Górny 2018].

Viruses are probably the most common cause of infections spreading in buildings [Barker et al. 2001]. The presence of viruses in the air and on indoor surfaces has been confirmed by numerous scientific studies [La Rosa et al. 2013; Prussin et al. 2015; Stobnicka-Kupiec et al. 2018]. According to the available data, viruses present on hands can lead to contamination of another 5–14 touched objects [Barker et al. 2004]. Contamination with viruses of one or two frequently touched areas, e.g. in an office building, can cause their spread to 40–60% of the remaining areas within 2–4 hours [Gerba 2014].

Although individual virus molecules can exist independently in the air, they tend to aggregate quickly (e.g. with dust particles) and easily settle on surfaces due to the gravity. According to some data, the spread of viral molecules in a building depends more on the distribution of room pressure than on the value of the ventilation airflow.

The indoor sources of viruses, which are mostly people, are indicated as more important than the outdoor ones. However, although airborne transmission is the most common way of transferring viruses between infected and healthy individuals, ventilation or air-conditioning systems are increasingly being cited as a potential route of transmission in the indoor environment [Verreault et al. 2008].

In the case of water damage to buildings, flood waters or sewage from sewage systems are important sources of virus particles. The risk of viral infections in flood-affected areas is an important health problem, and infectious diseases can cause outbreaks within weeks after the flooding occurs. The presence of enteroviruses, noroviruses, astroviruses, rotaviruses, adenoviruses and parechoviruses, as well as hepatitis A and E viruses, was recorded in municipal wastewater. Researchers also point to the possibility of influenza viruses, coronaviruses, poliomaviruses, human papillomaviruses and polio virus being present in wastewater [La Rosa et al. 2012; Lago et al. 2003].

## REFERENCES

Adan, O. C. G. 1994. *On the Fungal Defacement of Interior Finishes*. Eindhoven: Technical University.

Adhikari, A., J. Jung, T. Reponen et al. 2009. Aerosolization of fungi, (1–3)-β-D glucan, and endotoxin from flood-affected materials collected in New Orleans homes. *Environ Res* 109(3):215–224.

Anderson, A. S., and E. M. H. Wellington. 2001. The taxonomy of Streptomyces and related genera. *Int J Syst Evol Microbiol* 51(3):797–814.

Andersson, M. A., M. Nikulin, and U. Köljalg 1997. Bacteria, molds, and toxins in water-damaged building materials. *Appl Environ Microbiol* 63(2):387–393.

Aukrust, L., S. M. Borch, and R. Einarsson. 1985. Mold allergy: Spores and mycelium as allergen sources. *Allergy* 40:43–48.

Barker, J., D. Stevens, and S. F. Bloomfield. 2001. Spread and prevention of some common viral infections in community facilities and domestic homes. *J Appl Microbiol* 91(1):7–21.

Barker, J., I. B. Vipond, and S. F. Bloomfield. 2004. Effects of cleaning and disinfection in reducing the spread of Norovirus contamination via environmental surfaces. *J Hosp Infect* 58(1):42–44.

Beguin, H. 1995. Mould biodiversity in homes II: Analysis of mattress dust. *Aerobiologia* 11(1):3–10.

Beguin, H., and N. Nolard. 1994. Mould biodiversity in homes I: Air and surface analysis of 130 dwellings. *Aerobiologia* 10(2–3):157–166.

Bisht, V., B. P. Singh, S. N. Gaur, N. Arora, and S. Sridhara. 2000. Allergens of Epicoccum nigrum grown in different media for quality source material. *Allergy* 55(3):274–280.

Bloom, E., L. F. Grimsley, C. Pehrson, J. Lewis, and L. Larsson. 2009. Molds and mycotoxins in dust from water-damaged homes in New Orleans after Hurricane Katrina. *Indoor Air* 19(2):153–158.

Brasel, T. L., J. M. Martin, C. G. Carriker, S. C. Wilson, and D. C. Straus. 2005. Detection of airborne Stachybotrys chartarum macrocyclic trichothecene mycotoxins in the indoor environment. *Appl Environ Microbiol* 71(11):7376–7388.

Charpin-Kadouch, C., G. Maurel, and R. Felipo 2006. Mycotoxin identification in moldy dwellings. *J Appl Toxicol* 26(6):475–479.

Chełkowski, J. 1985. *Mikotoksyny, wytwarzające je grzyby i mikotoksykozy (Mycotoxins, fungi producing them and mycotoxicoses)*. Warsaw: Wyd. SGGW-AR.

Chew, G. L., J. Wilson, F. A. Rabito, et al. 2006. Mold and endotoxin levels in the aftermath of Hurricane Katrina: A pilot project of homes in New Orleans undergoing renovation. *Environ Health Perspect* 114(12):1883–1889.

Cole, E. C., K. K. Foarde, K. E. Leese, D. A. Green, D. L. Franke, and M. A. Berry. 1994. Assessment of fungi in carpeted environment. Air Quality Monographs. In *Vol. 2: Health Implications of Fungi in Indoor Environments*, eds. R. A. Samson, B. Flannigan, M. E. Flannigan, A. P. Verhoeff, O. C. G. Adan, and E. S. Hoekstra, 103–128. Amsterdam: Elsevier Science B.V.

Cyprowski, M., A. Ławniczek-Wałczyk, and R. L. Górny. 2016. Airborne peptidoglycans as a supporting indicator of bacterial contamination in a metal processing plant. *Int J Occup Med Environ Health* 29(3):427–437.

Cyprowski, M., A. Stobnicka-Kupiec, R. L. Górny, M. Gołofit-Szymczak, A. Ptak-Chmielewska, and A. Ławniczek-Wałczyk. 2019. Across-shift changes in upper airways after exposure to bacterial cell wall components. *Ann Agric Environ Med* 26(2):236–241.

Davis, P. J. 2001. Molds, toxic molds, and indoor air quality. *CRB Note* 8:1–16.

De Hoog, G. S., and J. Guarro, eds. 1995. *Atlas of Clinical Fungi*. Baarn-Delft: Centralbureau voor Schimmelcultures.

Douwes, J. 2005. (1–3)-β-D-glucans and respiratory health: A review of the scientific evidence. *Indoor Air* 15(3):160–169.

Dutkiewicz, J., L. Jabłoński, and S. A. Olenchock. 1988. Occupational biohazards: A review. *Am J Ind Med* 14(5):605–623.

Ensign, J. C. 1978. Formation, properties, and germination of actinomycete spores. *Annu Rev Microbiol* 32:185–219.

Esch, R. E. 2004. Manufacturing and standardizing fungal allergen products. *J Allergy Clin Immunol* 113(2):210–215.

Fadel, R., B. David, S. Paris, and J. L. Guesdon. 1992. Alternaria spore and mycelium sensitivity in allergic patients: In vivo and in vitro studies. *Ann Allergy* 69(4):329–335.

Fisher, F., and N. B. Cook. 1998. *Fundamentals of Diagnostic Mycology*. Philadelphia: W. B. Saunders Company.

Fisher, G., T. Müller, R. Schwalbe, R. Ostrowski, and W. Dott. 2000. Species-specific profiles of mycotoxins produced in cultures and associated with conidia of airborne fungi derived from biowaste. *Int J Hyg Environ Health* 203(2):105–116.

Flannigan, B., and J. D. Miller. 2001. Microbial growth in indoor environments. In *Microorganisms in Home and Indoor Work Environments*, eds. B. Flannigan, R. Samson, and J. D. Miller, 35–67. London: Taylor and Francis.

Gerba, C. P. 2014. Impact of a Quaternary Ammonium Compound (QAC) disinfectant on the spread of viruses in facilities. In *Proceedings of the 54th Interscience Conference on Antimicrobial Agents and Chemotherapy (ICAAC)*, Washington, 5–9th September 2014.

Gołofit-Szymczak, M., and R. L. Górny. 2018. Microbiological air quality in office buildings equipped with different ventilation systems. *Indoor Air* 28(6):792–805.

Góra, A., B. Mackiewicz, P. Krawczyk et al. 2009. Occupational exposure to organic dust, microorganisms, endotoxin and peptidoglycan among plants processing workers in Poland. *Ann Agric Environ Med* 16(1):143–150.

Górny, R. L. 1998. *Ocena właściwości aerozoli ziarnistych i bioaerozoli w mieszkaniach konurbacji górnośląskiej*. [Evaluation of properties of granular aerosols and bioaerosols in apartments of the Upper Silesian conurbation]. Sosnowiec: Śląska Akademia Medyczna.

Górny, R. L., and J. Dutkiewicz. 2002. Bacterial and fungal aerosols in indoor environment in Central and Eastern European countries. *Ann Agric Environ Med* 9(1):17–23.

Gots, R. E., N. J. Layton, and S. W. Pirages. 2003. Indoor health: Background levels of fungi. *Am Ind Hyg Assoc J* 64(4):427–438.
Gottschalk, C., J. Bauer, and K. Meyer. 2008. Detection of satratoxin G and H in indoor air from a water-damaged building. *Mycopathologia* 166(2):103–107.
Grant, C., C. A. Hunter, B. Flannigan, and A. F. Bravery. 1989. The moisture requirements of moulds isolated from domestic dwellings. *Int Biodeterior* 25(4):259–289.
Gravesen, S., P. A. Nielsen, R. Iversen, and K. F. Nielsen. 1999. Microfungal contamination of damp buildings: Examples of risk constructions and risk materials. *Environ Health Perspect* 107:505–508.
Heczko, P. B., M. Wróblewska, and A. Pietrzyk. 2014. *Mikrobiologia Lekarska.* [Medical Microbiology]. Warsaw: PZWL, 173–185.
Helbling, A., and A. Reimers. 2003. Immunotherapy in fungal allergy. *Curr Allergy Asthma Rep* 3(5):447–453.
Herbarth, O., U. Schlink, A. Müller, and M. Richter. 2003. Spatiotemporal distribution of airborne mould spores in apartments. *Mycol Res* 107(Pt 11):1361–1371.
Hiipakka, D. W., and J. R. Buffington. 2000. Resolution of sick building syndrome in a high-security facility. *Appl Occup Environ Hyg* 15(8):635–643.
Hirsch, S. R., and J. A. Sosman. 1976. A one-year survey of mould growth inside twelve homes. *Ann Allergy* 36(1):30–38.
Hoffman, D. R., P. P. Kozak, S. A. Gillman, L. H. Cummins, and J. Gallup. 1981. Isolation of spore-specific allergens from Alternaria. *Ann Allergy* 46(6):310–316.
Hoppe, K. A., N. Metwali, S. S. Perry, T. Hart, P. A. Kostle, and P. S. Thorne. 2012. Assessment of airborne exposures and health in flooded homes undergoing renovation. *Indoor Air* 22(6):446–456.
Horner, W.E., A. Helbling, J. E. Salvaggio, and S. B. Lehrer. 1995. Fungal allergens. *Clin Microbiol Rev* 8(2):161–179.
Hunter, C. A., C. Grant, B. Flannigan, and A. F. Bravery. 1988. Mould in buildings: The air spora of domestic dwellings. *Int Biodeterior* 24(2):81–101.
Hyvärinen, A., T. Meklin, A. Vepsäläinen, and A. Nevalainen. 2002. Fungi and Actinobacteria in moisture-damaged building materials: Concentrations and diversity. *Int Biodeterior Biodegrad* 49(1):27–37.
Hyvärinen, A., T. Reponen, T. Husman, J. Ruuskanen, and A. Nevalainen. 1993. Characterizing mold problem buildings: Concentrations and flora of viable fungi. *Indoor Air* 3(4):337–343.
Jolie, R., L. Bäckström, and P. Gunderson. 1998. Airborne contaminants and farmers health in swine farms with high and low prevalence of respiratory diseases in pigs. *Ann Agric Environ Med* 5(1):87–92.
Keller, R., K. Senkpiel., and H. Ohgke. 1998. Geruch Als Indicator fur Schimmelpilzbelastungen in Naturlich Belufteten Innenraumen: Nachweis mit Analytischer MVOC-Messung. Gesundheitliche Gefahren Durch Biogene Luftschadstoffe-Schriftenreihe Des Institues fur Medizinische Mikrobiologie un Hygiene Heft 2, Medizinische Universitat, and Lubeck. *Arch Microbiol* 179(2):75–82.
Kemp, P. C., H. G. Neumeister-Kemp, B. Esposito, G. Lysek, and F. Murray. 2003. Changes in airborne fungi from the outdoors to indoor air: Large HVAC systems in nonproblem buildings in two different climates. *Am Ind Hyg Assoc J* 64(2):269–275.
Kirk, P. M., P. F. Cannon, D. W. Minter, and J. A. Stalpers. 2011. *Dictionary of the Fungi*, 10th ed. Egham: CABI.
Korpi, A., J. Järnberg, and A. L. Pasanen. 2009. Microbial volatile organic compounds. *Crit Rev Toxicol* 39(2):139–193.
Kurup, V. P., and B. Banerjee. 2000. Fungal allergens and peptide epitopes. *Peptides* 21(4):589–599.

Kuske, M., A. C. Romain, and J. Nicolas. 2005. Microbial volatile organic compounds as indicators of fungi: Can an electronic nose detect fungi in indoor environments? *Build Environ* 40(6):824–831.

La Rosa, G., M. Fratini, S. Della Libera, M. Iaconelli, and M. Muscillo. 2012. Emerging and potentially emerging viruses in water environments. *Ann Ist Super Sanita* 48(4):397–406.

La Rosa, G., M. Fratini, S. Della Libera, M. Iaconelli, and M. Muscillo. 2013. Viral infections acquired indoors through airborne, droplet or contact transmission. *Ann Ist Super Sanita* 49(2):124–132.

Lacey, J. 1981. Airborne Actinomycete spores as respiratory allergens. *Zbl Bakt Suppl* 11:243–250.

Lacey, J. 1988. Actinomycetes as biodeteriogens and pollutants of the environment. In *Actinomycetes in Biotechnology*, eds. M. Goodfellow, S. T. Williams, and M. Mordarski, 359–432. San Diego: Academic Press, San Diego.

Lago, P. M., H. E. Gary, L. S. Pérez et al. 2003. Poliovirus detection in wastewater and stools following an immunization campaign in Havana, Cuba. *Int J Epidemiol* 32(5):772–777.

Laitinen, S., J. Kangas, K. Husman, and P. Susitaival. 2001. Evaluation of exposure to airborne bacterial endotoxins and peptidoglycans in selected work environments. *Ann Agric Environ Med* 8(2):213–219.

Larsen, L. 1994. Fungal allergens. In. *Air Quality Monographs, Vol. 2: Health Implications of Fungi in Indoor Environments*, eds. R. A. Samson, B. Flannigan, M. E. Flannigan, A. P. Verhoeff, O. C. G. Adan, and E. S. Hoekstra, 215–220. Amsterdam: Elsevier Science B.V.

Ławniczek-Wałczyk, A., and R. L. Górny. 2010. Endotoxins and β-glucans as markers of microbiological contamination: Characteristics, detection, and environmental exposure. *Ann Agric Environ Med* 17(2):193–208.

Lehtonen, M., T. Reponen, and A. Nevalainen. 1993. Everyday activities and variation of spore concentration in indoor air. *Int Biodeterior Biodegrad* 31(1):25–39.

Licorish, K., H. S. Novey, P. Kozak, R. D. Fairshter, and A. F. Wilson. 1985. Role of Alternaria and Penicillium spores in the pathogenesis of asthma. *J Allergy Clin Immunol* 76(6):819–825.

Liebers, V., M. Raulf-Heimsoth, and T. Brüning. 2008. Health effects due to endotoxin inhalation (review). *Arch Toxicol* 82:203–210.

Lighthart, B., and L. D. Stetzenbach. 1994. Distribution of microbial bioaerosol. In *Atmospheric Microbial Aerosols: Theory and Applications*, eds. B. Lighthart, and A. J. Mohr, 68–98. New York: Chapman and Hall, Inc.

Mandrioli, P., G. Caneva, and C. Sabbioni, eds. 2003. *Cultural Heritage and Aerobiology*. Dordrecht: Kluwer Academic Publishers.

Matakakis, T. Z., C. Barnes, and E. R. Tovey. 2001. Spore germination increases allergen release from Alternaria. *J Allergy Clin Immunol* 107(2):388–390.

McBride, M. J., and J. C. Ensign. 1987. Effects of intracellular trehalose content on Streptomyces griseus spores. *J Bacteriol* 169(11):4995–5001.

Miller, J. D. 1992. Fungi as contaminants in indoor air. *Atmos Environ* 26(12):2163–2172.

Miller, J. D., A. M. Laflamme, Y. Sobol, P. Lafontaine, and R. Greenhalgh. 1988. Fungi and fungal products in some Canadian houses. *Int Biodeterior* 24(2):103–120.

Morey, P. R. 1999. Comparison of airborne culturable fungi in moldy and non-moldy buildings. *Proc of Indoor Air* 99(2)2:524–528.

Myhre, A. E., A. O. Aasen, C. Thiemermann, and J. E. Wang. 2006. Peptidoglycan: An endotoxin in its own right? *Shock* 25(3):227–235.

Nevalainen, A., A. L. Pasanen, M. Niininen, T. Reponen, P. Kalliokoski, and M. J. Jantunen. 1991. The indoor air quality in Finnish homes with mold problems. *Environ Int* 17(4):299–302.

Nielsen, K. F. 2003. Mycotoxin production by indoor molds. *Fungal Genet Biol* 39(2):103–117.
Paris, S., C. Fitting, E. Ramires, J. P. Latgé, and B. David. 1990b. Comparison of different extraction methods of Alternaria allergens. *J Allergy Clin Immunol* 85(5):941–948.
Pasanen, A. L., M. Niininen, P. Kalliokoski, A. Nevalainen, and M. J. Jantunen. 1992a. Airborne Cladosporium and other fungi in damp versus reference residences. *Atmos Environ* 26(1):121–124.
Pasanen, A. L., T. Juutinen, M. J. Jantunen, and P. Kalliokoski. 1992b. Occurrence and moisture requirements of microbial growth in building materials. *Int Biodeterior Biodegrad* 30(4):273–283.
Portnoy, J., F. Pachero, Y. Ballam, and C. Barnes. 1993. The effect of time and extraction buffers on residual protein and allergen content of extracts derived from four strains of Alternaria. *J Allergy Clin Immunol* 91(4):930–938.
Prussin, A., E. B. Garcia, and L. C. Marr. 2015. Total concentrations of virus and bacteria in indoor and outdoor air. *Environ Sci Technol Lett* 2(4):84–88.
Rando, R. J., C. W. Kwon, and J. J. Lefante. 2014. Exposures to thoracic particulate matter, endotoxin, and glucan during post-Hurricane Katrina restoration work, New Orleans 2005–2012. *J Occup Environ Hyg* 11(1):9–18.
Rao, C. Y., M. A. Riggs, G. L. Chew et al. 2007. Characterization of airborne molds, endotoxins, and glucans in homes in New Orleans after Hurricanes Katrina and Rita. *Appl Environ Microbiol* 73(5):1630–1634.
Reijula, K. E., V. P. Kurup, and J. N. Fink. 1991. Ultrastructural demonstration of specific IgG and IgE antibodies binding to Aspergillus fumigatus from patients with aspergillosis. *J Allergy Clin Immunol* 87(3):683–688.
Reponen, T., A. Nevalainen, M. Jantunen, M. Pellikka, and P. Kalliokoski. 1992. Normal range criteria for indoor air bacteria and fungal spores in subarctic climate. *Indoor Air* 2(1):26–31.
Rylander, R., K. Persson, H. Goto, K. Yuasa, and S. Tanaka. 1992. Airborne beta-1,3-glucan may be related to symptoms in sick buildings. *Indoor Environ* 1(5):263–267.
Sebastian, A., A. M. Madsen, L. Martensson, D. Pomorska, and L. Larsson. 2006. Assessment of microbial exposure risks from handling of biofuel wood chips and straw-effect of outdoor storage. *Ann Agric Environ Med* 13(1):139–145.
Sigsgaard, T., E. C. Bonefeld-Jorgensen, H. J. Hoffmann, J. Bønløkke, and T. Krüger. 2005. Microbial cell wall agents as an occupational hazard. *Toxic Appl Pharm* 207:310–319.
Solomon, G. M., M. Hjelmroos-Koski, M. Rotkin-Ellman, and S. K. Hammond. 2006. Airborne mold and endotoxin concentrations in New Orleans, Louisiana, after flooding, October through November 2005. *Environ Health Perspect* 114(9):1381–1386.
Solomon, W. R., H. A. Burge, and L. Muilenberg. 1980. Allergenic properties of Alternaria spore, mycelium, and "metabolic" extracts. *J Allergy Clin Immunol* 65:229.
Soroka, P. M., M. Cyprowski, and I. Szadkowska-Stańczyk. 2008. Narażenie zawodowe na mykotoksyny w różnych gałęziach przemysłu. [Occupational exposure to mycotoxins in various industries]. *Med Pr* 59(4):333–345.
Sporik, R. B., L. K. Arruda, J. Woodfolk, M. D. Capman, and T. A. E. Platts-Mills. 1993. Environmental exposure to Aspergillus fumigatus allergen (Asp f 1). *Clin Exp Allergy* 23(4):326–331.
St-Germain, G., and R. Summerbell. 1996. *Identifying Filamentous Fungi*. Belmont: Star Publishing Company.
Stobnicka, A., M. Gołofit-Szymczak, A. Wójcik-Fatla, V. Zając, J. Korczyńska-Smolec, and R. L. Górny. 2018. Prevalence of human parainfluenza viruses and noroviruses genomes on office fomites. *Food Environ Virol* 10(2):133–140.
Stobnicka-Kupiec, A., and R. L. Górny. 2018. Metody detekcji wirusów w różnych środowiskach pracy. [Virus detection methods in different working environments]. *Podstawy i Metody Oceny Środowiska Pracy* 3(97):5–18.

Ström, G., J. West, B. Wessén, and U. Palmgren. 1994. Quantitative analysis of microbial volatiles in damp Swedish houses. In *Health Implication of Fungi in Indoor Air Environments*, eds. R. A. Samson, B. Flannigan, and M. E. Flannigan. Amsterdam: Elsevier.

Sunesson, A. L., C. A. Nilsson, R. Carlson, G. Blomquist, and B. Andersson. 1997. Production of volatile metabolites from Streptomyces albidoflavus cultivated on gypsum board and tryptone glucose extract agar: Influence of temperature, oxygen and carbon dioxide levels. *Ann Occup Hyg* 41(4):393–413.

Van Reenen-Hoekstra, E. S., R. A. Samson, A. P. Verhoeff, J. H. van Wijnen, and B. Brunekreef. 1991. Detection and identification of moulds in Dutch houses and non-industrial working environments. *Grana* 30(2):418–423.

Verreault, D., S. Moineau, and C. Duchaine. 2008. Methods for sampling of airborne viruses. *Microbiol Mol Biol Rev* 72(3):413–444.

Watkinson, S. C., L. Boddy, and N. P. Money, eds. 2015. *The Fungi*, 3rd ed. London: Academic Press.

Wilkins, K., K. Larsen, and M. Simkus. 2000. Volatile metabolites from mold growth on building materials and synthetic media. *Chemosphere* 41(3):437–446.

Williams, S. T., M. E. Sharpe, and J. G. Holt, eds. 1989. *Bergey's Manual of Systematic Bacteriology*, Vol. 4. Baltimore: Williams and Wilkins.

Zaremba, M. L., and J. Borowski. 2001. *Mikrobiologia Lekarska*. Warsaw: Wydawnictwo Lekarskie PZWL.

# 3 Epidemiology of Microbiological Contamination of Indoor Environments

*Anna Ławniczek-Wałczyk*

## CONTENTS



Microorganisms and dampness have an important impact on both the technical conditions of a building and the health of its occupants and as such they pose a serious threat to public health and the economy in many countries. The data from 31 European countries show that about 10–15% of buildings are affected by dampness and contaminated by moulds [Haverinen-Shaughnessy 2012]. In the US, about 18% of homes have experienced damage caused by flooding, and more than half of them have been contaminated by mould [Mudarri and Fisk 2007]. It is estimated that living in damp houses affected by bio-corrosion of construction materials increases the risk of asthma and other adverse respiratory health effects by 30% to 50% [Fisk et al. 2007]. A study conducted in 2017 shows that 1 in 6 Europeans (and in some countries even 1 in 3) experiences problems with excessive dampness and moulds, while 1.5 times more people living in sick buildings have health problems, compared to people living in healthy ones [Rasmussen et al. 2017]. Microorganisms that cause biological corrosion of construction and finishing materials in buildings may adversely affect the occupants by releasing a range of harmful structures and substances into the air [IOM 2004; Hyvärinen et al. 1993; Hope 2013; Adhikari et al. 2009]. Economic consequences of illnesses associated with exposure to dampness and moulds are significant. It is estimated that in Europe, the costs of asthma and chronic obstructive pulmonary disease treatment amount to €82 billion annually.

Half of this amount represents direct costs, including medication and healthcare expenses, while the other half are indirect costs, related to absence or reduced work productivity [Rasmussen et al. 2017]. In the US, it is estimated that about 15–20% of economic costs of adverse health effects in the population are related to the exposure to dampness and mould contamination in private houses and public buildings. Average annual costs of the associated treatment are estimated at $15.1 billion for asthma, $3.7 for allergic rhinitis, $1.9 billion for bronchitis and $1.7 billion for deaths resulting from asthma [Mudarri 2016].

## 3.1 ENVIRONMENTAL PREVALENCE AND SPREADING OF MICROBIOLOGICAL CONTAMINANTS

The main source of microorganisms (especially moulds, actinomycetes, Gram-positive cocci and rods, including bacilli) in buildings is the outdoor environment, from which they can enter the buildings through doors, windows, vents and heating or air-conditioning systems. These microorganisms may also be carried into the building on clothing and footwear and by pets. An active reservoir of microorganisms may also be the users themselves as well as plants and different items and materials stored indoors [Lehtonen et al. 1993; Zyska 1999; Flannigan and Miller 2011]. It should be remembered that microorganisms first colonise spaces with increased humidity (kitchens, bathrooms, basements, attics, floors, spaces near windows etc.). Moulds, as saprophytes, grow very well on materials containing cellulose, i.e. on wood, cardboard, paper products, wallpapers, ceiling panels and many other furnishing products. They may colonise surfaces covered with paints and adhesives, grow on gypsum boards and in dust deposited in cracks of floors, walls and ceilings, as well as on textiles [Zyska 1999; Spengler et al. 2000; Flannigan and Miller 2011]. It should always be remembered not to carry materials and other furnishings that show visible signs of bio-corrosion into the building as they may be a secondary source of microbiological contamination [IOM 2004; Adhikari et al. 2009; Balasubramanian et al. 2012].

Microorganisms present in buildings affected by bio-corrosion may influence the occupants' health in many ways. Inhalation is the most common route of exposure to moulds and bacteria; however, exposure may also occur as a result of direct contact with microorganisms through damaged skin or mucous membranes or using the oral route by transferring microbial pollutants with dirty hands, for example. A variety of harmful microbiological agents occurring in a sick building may cause extreme difficulty in determining the exact reason for a health problem connected with occupying contaminated spaces (Figure 3.1). Such a situation from both medical and epidemiological points of view makes precise estimation of the number of adverse cases (directly related to microbiological contamination of premises or resulting from the presence of water damage) very complicated.

## 3.2 HEALTH SIGNIFICANCE OF MICROBIOLOGICAL HAZARDS

### 3.2.1 Allergic Reactions

Microorganisms colonising water-damaged buildings play an important role in the pathogenesis of many allergic diseases, including asthma, allergic alveolitis, allergic

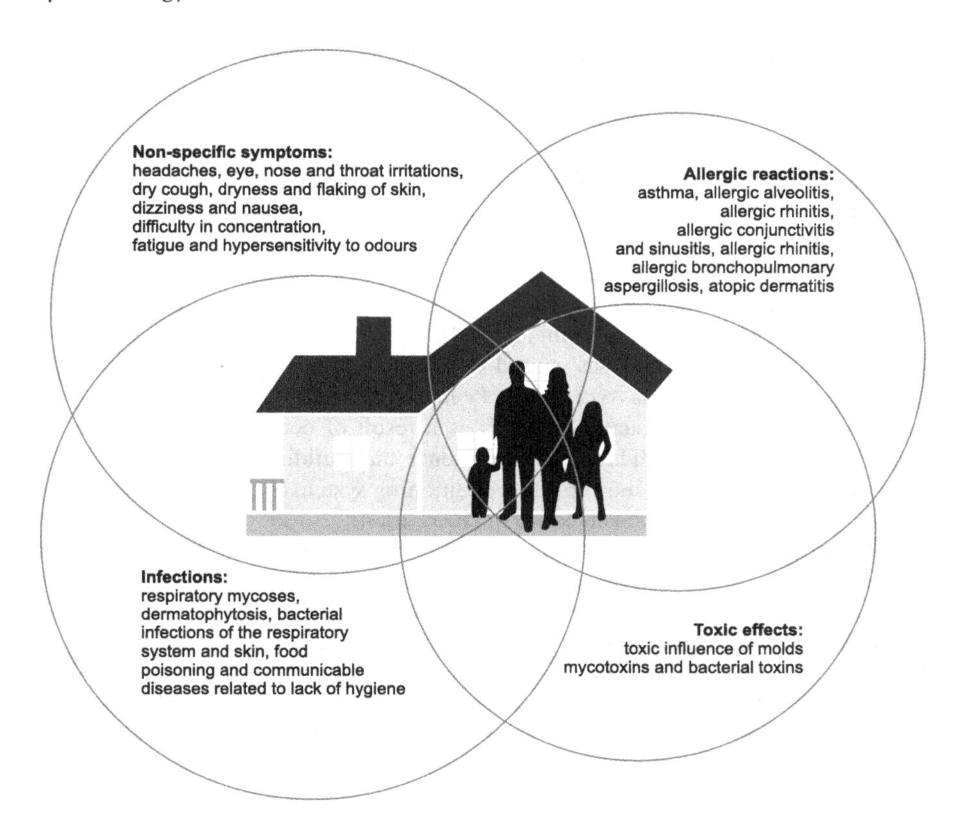


**FIGURE 3.1** Health effects and diseases resulting from indoor dampness and mould exposure.

rhinitis, allergic conjunctivitis and sinusitis, allergic inflammation of the nasal mucosa, sarcoidosis and allergic bronchopulmonary aspergillosis. Allergic reactions to moulds are quite common and can affect about 20% of the world's population (3–10% in Europe). The most allergenic fungi include those of *Alternaria*, *Aspergillus*, *Cladosporium*, *Mucor*, *Penicillium* and *Trichoderma* genera [Górny and Dutkiewicz 2002; Denning et al. 2014; Kurup et al. 2002; Hurraß et al. 2017]. Hypersensitivity to fungal allergens is a significant risk factor for the development of severe bronchial asthma. It is estimated that, among 22 million people with asthma, about 5 million cases are caused by poor living conditions, in particular increased humidity and associated exposure to moulds [Mudarri and Fisk 2007]. According to the 'Healthy Homes Barometer 2017' report, about 2.2 million Europeans suffer from asthma caused by their living conditions [Rasmussen et al. 2017].

The latest studies indicate the relationship between exposure to moulds in early childhood and the development of atopic diseases, including asthma, later in life. Increased humidity and visible traces of moulds or their odours in the building are considered to be determinants of the development of asthma and other respiratory or skin diseases in exposed individuals [Platt et al. 1989; Koskinen et al. 1995; Meklin et al. 2002; Quansah et al. 2012; Karvonen et al. 2015; Oluwole et al. 2016; Moses et al. 2019; Wang et al. 2019]. The results of a meta-analysis conducted by Sharpe et al. [2015] suggest that the presence of fungi in the air, in particular of *Aspergillus*, *Penicillium*, *Cladosporium* and *Alternaria* genera, exacerbates the existing symptoms of asthma in children and adults. In some patients with asthma, the inhaled conidia *Aspergillus fumigatus* (commonly found in damp houses) are not effectively eliminated from the airways and form colonies growing in the bronchial lumen. This may lead to the development of allergic bronchopulmonary aspergillosis (ABPA), a very serious disease caused by hypersensitivity to the antigens of this fungus. In a medical history, ABPA is very rarely diagnosed without bronchial asthma, which results in the lack of data on the incidence of the disease in the general population [Denning et al. 2014; Shah and Panjabi 2014]. In turn, hypersensitivity pneumonitis (HP) is a complex of diseases caused by inhalation exposure to microbial antigens including fungi such as *Aspergillus*, *Penicillium*, *Cladosporium*, *Trichosporon* and *Aureobasidium*. HP usually develops as a result of occupational exposure to fungi [Hurraß et al. 2017]. There is also evidence that buildings contaminated with fungi and their equipment (showers, air-conditioning systems or humidifiers) can be sources of the antigens responsible for HP [IOM 2004].

Actinomycetes and Gram-positive rods frequently occurring in damp buildings may also pose a significant health risk to exposed individuals. It has been shown that the inhalation of actinomycete spores of *Streptomyces albus* can cause acute pulmonary disease and allergic reactions (including allergic alveolitis). Exposure to thermophilic actinomycetes and nontuberculosis mycobacteria occurring in damp buildings may manifest itself in decreased lung function, asthma and numerous respiratory and systemic symptoms. These relationships proved to be epidemiologically stronger when simultaneous exposure to endotoxins and moulds was taken into account [Kagen 1981; Park et al. 2017].

Although evidence that building dampness and the associated development of moulds and other microorganisms are closely related to adverse respiratory effects has been abundantly accumulated in scientific literature, the relationship between the concentration of microorganisms and their responsibility for the development of respiratory symptoms has not been fully determined and still remains controversial in many aspects [Górny et al. 2011; Sharpe et al. 2015; Sheehan and Phipatanakul 2016; Hurraß et al. 2017].

### 3.2.2 Toxic Reactions

Microorganisms colonising damp buildings and construction materials may cause irritation, inflammation and toxic reactions of varying severity in exposed individuals. Among the best-known toxins are mycotoxins produced by moulds. Mycotoxins are harmful for human health, especially when they enter the body through the oral route [Zain 2011]. The spectrum of their effects is very broad: from mutagenic, through teratogenic, cytotoxic, nephrotoxic and hepatotoxic effects, to neurotoxic ones. Skin contact with objects contaminated with mycotoxins may also be a source of exposure. *In vitro* studies have shown that aflatoxin B1, ochratoxin, citrinin, T2 toxin or zearalenone may penetrate through the skin quite freely [Hope 2013]. Unfortunately, little is known still about the causative role of mycotoxins in the development of respiratory diseases. However, it has been found that inhalation of fungal toxins may lead to the impairment of neuromotor functions in the airways. Inhalation of aflatoxin-contaminated dust creates a risk of liver, trachea, lung and bronchial cancers [Yang and Johanning 1997]. Mycotoxin contamination of the indoor environment is a relatively new problem, which gained attention after 1994, when the American Center for Disease Control and Prevention (CDC) described several cases of primary idiopathic pulmonary hemosiderosis in children in Ohio, USA, developed after exposure to *Stachybotrys chartarum* fungus (which requires very wet conditions for growth) and its toxins [Bennett and Klich 2003]. The study by Jarvis et al. [1998] on the cytotoxicity of fungi, including *S. chartarum* and *Memoniella* spp. isolated from mouldy residential buildings, showed that mycotoxins produced by these fungi, such as satratoxin, trichoverrol and roridin, were toxic to lung epithelial cells. Other studies have confirmed that occupying damp and mouldy buildings may be connected with exposure to mycotoxins, including ochratoxins, aflatoxins and trichothecenes [Hooper et al. 2009; Thrasher et al. 2012; Hope 2013]. In the study by Thrasher et al. [2012], residents of a mouldy house complained of various health problems, such as chronic sinusitis, neurologic deficits, wheezing coughs, spots on the skin, skin redness, nosebleeds, fatigue and many other non-specific symptoms. Subsequent detailed evaluation of the premises revealed that mycotoxins, including trichothecenes, aflatoxins and ochratoxins were found in the samples of tissues and body fluids (urine, nasal discharge, breast milk, placenta and umbilical cord) collected from the residents and in the samples of construction and finishing materials.

Until now, there have been no standardized diagnostic methods enabling rapid identification of diseases caused by exposure to mycotoxin-producing moulds. However, some studies are being carried out on the use of immunological tests to identify mycotoxins in serum, tissues and body fluids of individuals exposed to moulds. The results obtained have revealed that urine samples seem to be the best material for screening tests [Hooper et al. 2009].

Actinomycetes growing on damp or water-damaged building materials may also produce a range of toxins (e.g. valinomycin) [Andersson et al. 1998]. *In vitro* studies show that actinomycete spores from *Streptomyces* genus, isolated from mouldy buildings, are strong stimulators of macrophages in mouse and human cell lines, inducing production of proinflammatory mediators. Actinomycete spores may be even more active in this respect than fungal conidia [Huttunen et al. 2003]. However, it should be clearly stated that data on human health effects are limited and the exposure assessment still remains very difficult.

Inhalation of bacterial endotoxin may cause adverse health effects of inflammatory and toxic nature in exposed individuals [Balasubramanian et al. 2012; Rando et al. 2014]. The toxicity of endotoxin is manifested in a range of pathophysiological reactions which depend on the bacterial strain from which the endotoxin is derived. Typical reactions of the human body to endotoxins include elevated body temperature, flu-like symptoms, coughing and decreased blood pressure [Ławniczek-Wałczyk and Górny 2010]. Inhaled endotoxins activate pulmonary macrophages, which secrete various chemical compounds (e.g. cytotoxins) responsible for the development and course of the inflammatory reaction, resulting in acute inflammation, bronchoconstriction and disruption of gas exchange in the peripheral parts of the lungs. The effect of LPS can be local and/or generalised, leading to carbohydrate and fat metabolism disorders and vascular diseases, which may often end in shock and death. The adverse effects of endotoxin on the human body depend on the duration of exposure, its concentration, depth of penetration and respiratory deposition [Rylander and Holt 1998; Liebers et al. 2008; Mackiewicz 2013]. Occupational exposure of workers to endotoxins, (1→3)-β-D-glucans and other microbial agents in a dusty environment may result in organic dust toxic syndrome (ODTS), also known as inhalation fever or pulmonary mycotoxicosis. The syndrome may manifest itself in fever with chills, dry throat, headaches, osteoarticular pain and tightness of the chest or general malaise. These symptoms are usually transient and appear during contact with organic dust or shortly after exposure and usually disappear spontaneously [Rylander 1994]. This disease does not require any specific treatment. It is believed that chronic bronchitis may be a distant consequence of ODTS. Most ODTS cases are found among workers exposed to grain dust or dust produced from animal husbandry [Mackiewicz 2013].

### 3.2.3 Infections

Massive water damage is always connected with epidemiological risks. Poor hygienic and sanitary conditions during flooding and directly after the water has subsided may lead to an increased incidence of: communicable diseases caused by bacteria of *Vibrio* (cholera), *Rickettsia prowazekii* (typhus) and *Leptospira* (leptospirosis) genera, *Legionella pneumophila* (atypical pneumonia and Pontiac fever), nontuberculous *Mycobacterium avium* complex (mycobacteriosis) or type A viruses (hepatitis A); food poisoning caused by bacteria including *Salmonella*, *Shigella*, *Escherichia coli*, *Staphylococcus aureus*, *Bacillus* and *Campylobacter*, as well as noroviruses and rotaviruses; malaria caused by protozoa of *Plasmodium* genus; yellow fever caused by viruses of *Flavivirus* genus or the West Nile fever caused by a virus of the same name as well as dermatosis and other diseases such as acute respiratory infections, conjunctivitis and ear and throat infections [Zyska 1999; Brown and Murray 2013].

Long-term effects of water damage to buildings are often more severe for their users than the flooding itself. Once the water has subsided from the flooded area, harmful microbiological agents develop, in particular those of fungal and bacterial origin. It has been found that performing remedial works in buildings damaged by water and with visible signs of bio-corrosion exposes residents, workers and volunteers to harmful microbiological agents [Rando et al. 2014]. More than 43% of those working on repairing homes damaged by flood after Hurricane Katrina were found to have skin infections. It has also been demonstrated that the risk of skin problems increased more than 20 times among persons staying overnight in such buildings [Noe et al. 2007]. However, it should be emphasised that most of the microorganisms found in damp buildings belong to risk group 1 or 2 according to the classification in 'Directive 2000/54/EC on the protection of workers from risks related to exposure to biological agents at work'[Directive 2000/54/EC]. These are typically saprophytic microorganisms or opportunistic pathogens, which may cause diseases in humans and may be dangerous for workers, but their spread in the human population is unlikely and there are usually effective methods of their prevention and treatment.

The most important infectious agents found in damp buildings include moulds. The results of meta-analysis conducted by Sauni et al. [2013] indicate that the removal of damage caused by water and the disinfection of mouldy residential and office buildings result in significant alleviation of asthma symptoms and decrease in the incidence of respiratory infections in adults (e.g. flu-like symptoms, inflammation of the nasal mucosa, tonsillitis, otitis media, bronchitis, sinusitis, pneumonia). It should be emphasised that most infections caused by moulds develop in immunocompromised persons and are opportunistic. Particularly vulnerable are individuals with respiratory diseases (asthma, cystic fibrosis), infants and children, the elderly and those with compromised immune systems, due, among others, to cancer, acute leukaemia and human immunodeficiency virus (HIV) infection and acquired immune deficiency syndrome (AIDS) [Bongomin et al. 2017; Hurraß et al. 2017]. Only a few types of fungi are fully pathogenic (e.g. strains from *Histoplasma* and *Blastomyces* genera) and cause infections in humans with a compromised immune system. The most common etiological agents of mycoses are species of *Aspergillus* and *Candida* genera. Infections caused by *Candida* yeasts are mostly endogenous, less often exogenous. In turn, mycoses caused by fungi of *Aspergillus* genus (*A. fumigatus*, *A. flavus*, *A. niger*, *A. nidulans*, *A. terreus*) are exogenous and most often the main route of exposure is contaminated air. The most common clinical form of the infection is invasive pulmonary aspergillosis and generalised aspergillosis, rarely sinusitis. It is worth mentioning that there are about 3 million people suffering from pulmonary aspergillosis in the world [Bongomin et al. 2017; Hurraß et al. 2017; Anderson et al. 2017].

In case of invasive infections caused by fungi and bacteria, it is observed that treatment becomes increasingly difficult due to the growing resistance of pathogens to antimicrobials. Alarming pathogens, particularly dangerous due to therapeutic limitations, are found among yeast (e.g. *Candida* species resistant to fluconazole or other drugs from the azole and candin groups), bacteria (e.g. methicillin-resistant *Staphylococcus aureus*, MRSA and vancomycin-resistant *Enterococcus* strains, VRE), which cause approximately 3.4 thousand, 80 thousand and 20 thousand instances of disease, respectively. According to the CDC report, drug-resistant pathogen infections affect more than 2 million US citizens each year, of which 23 thousand die [CDC 2013].

### 3.2.4 Other Non-Specific Symptoms

According to epidemiological data, persons working in or occupying (e.g. office) buildings for a long time, often complain of numerous health problems of a non-specific nature. It is estimated that about 89 million Americans work in an indoor environment, and the number of workers affected by these problems may be as high as 35–60 million [Mendell et al. 2002]. The most frequently reported complaints include: headaches, eye, nose and throat irritation, dry cough, dryness and flaking of skin, dizziness and nausea, difficulty in concentration, fatigue and hypersensitivity to odours. These ailments are usually more severe the longer the exposed persons stay indoors, and most of them disappear after leaving the building. This syndrome of non-specific, subjective symptoms that arise as a result of staying indoors is described as 'sick building syndrome (SBS)'.

SBS symptoms can be influenced by both individual (sex, age, health, smoking habits) and environmental factors (poor air quality, dampness and mould in buildings, volatile organic compounds, poor ventilation and heating parameters) [IOM 2004; Smedje et al. 2017]. Among microbiological factors, the most important are: fungal (1→3)-β-D-glucans, bacterial endotoxins and peptidoglycans, microbial volatile organic compounds (MVOCs) and mycotoxins. Rylander et al. [1992] were the first who noticed the relationship between high concentrations of (1→3)-β-D-glucans in the air and the occurrence of such symptoms as irritation to the eyes and throat, coughing or itchy skin indoors. Persons with health problems such as respiratory diseases or atopy are more sensitive to (1→3)-β-D-glucans than healthy persons [Rylander and Holt 1998; Ławniczek-Wałczyk and Górny 2010]. Prolonged exposure to endotoxins may be the cause of headache, joint pain, coughing, flu-like symptoms or dyspnoea [Liebers et al. 2008]. Persons occupying damp and mouldy indoor spaces may also experience symptoms associated with the presence of microbial volatile organic compounds [Fischer and Dott 2003; IOM 2004]. Exposure of residents to high concentrations of MVOCs in the air may cause mucous membrane irritation, fatigue, headaches and general malaise [Fischer and Dott 2003; Araki et al. 2012].

## 3.3 SMELL NUISANCE (ODOURS)

Occupants of damp buildings often complain about the presence of a 'musty' smell, which is most frequently caused by microbial volatile organic compounds. The human nose is able to detect even extremely low concentrations in the air. Therefore, many questionnaire studies use their prevalence as an indirect indicator or predictor of mould presence and excessive dampness in the building [IOM 2004; Araki et al. 2012; Meklin et al. 2002; Karvonen et al. 2015; Moses et al. 2019]. To this day, the role of MVOCs in causing adverse health effects in persons occupying damp buildings affected by bio-corrosion is still not fully understood. However, numerous, non-specific symptoms (such as fatigue, lowered concentration, headaches, nausea or insomnia) are the most common effects of exposure to MVOCs. Symptoms of irritation and toxic reactions that occur after exposure to unpleasant odours are highly variable and are still the subject of scientific and epidemiological research [Inamdar et al. 2014; Bennett and Inamdar 2015; Hurraß et al. 2017]. In a study conducted in Finland, it was found that children living in houses with a perceptible smell of moulds had an increased risk by over 100% of developing asthma within the next six years of life [Jaakkola et al. 2005]. By contrast, in a study conducted in the south-west of England on the influence of fungal contamination of residential buildings on the development of allergic diseases in elderly, an unpleasant odour of moulds was found in more than 80% of the inspected houses, while 45% of the houses had visible traces of microbiological contamination. Exposure to moulds and unpleasant odour was also shown to be a risk factor for the development of asthma in persons (especially women) over the age of 50 [Moses et al. 2019]. Taking these facts into consideration, it is important to emphasise that any action aimed at reducing the undesirable presence of microorganisms in the indoor environment will certainly help to decrease the prevalence of adverse health effects among their users.

## REFERENCES


Adhikari, A., J. Jung, T. Reponen et al. 2009. Aerosolization of fungi, (1–3)-β-D glucan, and endotoxin from flood-affected materials collected in New Orleans homes. *Environ Res* 109(3):215–224.

Anderson, S. E., C. Long, and G. S. Dotson. 2017. Occupational allergy. *Eur Med J (Chelmsf)* 2(2):65–71.

Andersson, M., R. Mikkola, R. M. Kroppenstedt et al. 1998. The mitochondrial toxin produced by Streptomyces griseus strains isolated from an indoor environment is valinomycin. *Appl Environ Microbiol* 64(12):4767–4773.

Araki, A., A. Kanazawa, T. Kawai et al. 2012. The relationship between exposure to microbial volatile organic compound and allergy prevalence in single-family homes. *Sci Total Environ* 423:18–26.

Balasubramanian, R., P. Nainar, and A. Rajasekar. 2012. Airborne bacteria, fungi, and endotoxin levels in residential microenvironments: A case study. *Aerobiologia* 28(3):375–390.

Bennett, J. W., and A. A. Inamdar. 2015. Are some fungal volatile organic compounds (VOCs) mycotoxins? *Toxins (Basel)* 7(9):3785–3804.

Bennett, J. W., and M. Klich. 2003. Mycotoxins. *Clin Microbiol Rev* 16(3):497–516.

Bongomin, F., S. Gago, R. O. Oladele, and D. W. Denning. 2017. Global and multi-national prevalence of fungal diseases-estimate precision. *J Fungi (Basel)* 3(4):57.

Brown, L., and V. Murray. 2013. Examining the relationship between infectious diseases and flooding in Europe: A systematic literature review and summary of possible public health interventions. *Disaster Health* 1(2):117–127.

CDC [Centers for Disease Control and Prevention]. 2013. Antibiotic resistance threats in the United States, Atlanta. https://www.cdc.gov/drugresistance/biggest_threats.html [accessed October 11, 2019].

Denning, D. W., C. Pashley, D. Hartl et al. 2014. Fungal allergy in asthma-state of the art and research needs. *Clin Transl Allergy* 4:14.

Directive 2000/54/EC of the European Parliament and of the council of 18 September 2000 on the protection of workers from risks related to exposure to biological agents at work. *Official Journal of European Communities L.* 262/21, Brussels (with subsequent amendments: Commission Directive (EU) 2019/1833 of 24 October 2019 amending Annexes I, III, V and VI to Directive. 2000/54/EC of the European Parliament and of the Council as Regards Purely Technical Adjustments. *Official Journal of European Communities L* 279/54).

Fischer, G., and W. Dott. 2003. Relevance of airborne fungi and their secondary metabolites for environmental, occupational, and indoor hygiene. *Arch Microbiol* 179(2):75–82.

Fisk, W. J., Q. Lei-Gomez, and M. J. Mendell. 2007. Meta-analyses of the associations of respiratory health effects with dampness and mold in homes. *Indoor Air* 17(4):284–296.

Flannigan, B., and J. D. Miller. 2011. Microorganisms in home and indoor work environments. In: *Microorganisms in Home and Indoor Work Environments Diversity, Health Impacts, Investigation and Control*, eds. B. Flannigan, R. A. Samson, and J. D. Miller, 49–145. London: Taylor and Francis.

Górny, R. L., and J. Dutkiewicz. 2002. Bacterial and fungal aerosols in indoor environment in Central and Eastern European countries. *Ann Agric Environ Med* 9(1):17–23.

Górny, R. L., M. Cyprowski, A. Ławniczek-Wałczyk, M. Gołofit-Szymczak, and L. Zapór. 2011. Biohazards in the indoor environment: A role for threshold limit values in exposure assessment. In: *Management of Indoor Air Quality*, ed. M. Dudzińska, 1–20. London: CRC Press.

Haverinen-Shaughnessy, U. 2012. Prevalence of dampness and mold in European housing stock. *J Expo Sci Environ Epidemiol* 22(5):461–467.

Hooper, D. G., V. E. Bolton, F. T. Guilford, and D. C. Straus. 2009. Mycotoxin detection in human samples from patients exposed to environmental molds. *Int J Mol Sci* 10(4):1465–1475.

Hope, J. 2013. A review of the mechanism of injury and treatment approaches for illness resulting from exposure to water-damaged buildings, mold and mycotoxins. *Sci World J* 2013:20.

Hurraß, J., B. Heinzow, U. Aurbach et al. 2017. Medical diagnostics for indoor mold exposure. *Int J Hyg Environ Health* 220(2 Pt B):305–328.

Huttunen, K., A. Hyvärinen, A. Nevalainen, H. Komulainen, and M. R. Hirvonen. 2003. Production of proinflammatory mediators by indoor air bacteria and fungal spores in mouse and human cell lines. *Environ Health Perspect* 111(1):85–92.

Hyvärinen, A., R. Reponen, T. Husman, J. Ruuskanen, and A. Nevalainen. 1993. Characterizing mold problem buildings: Concentrations and flora of viable fungi. *Indoor Air* 3(4):337–343.

Inamdar, A. A., T. Zaman, S. U. Morath, D. C. Pu, and J. W. Bennett. 2014. Drosophila melanogaster as a model to characterize fungal volatile organic compounds. *Environ Toxicol* 29(7):829–836.

IOM [Institute of Medicine, Committee on Damp Indoor Spaces and Health. Board of Health Promotion and Disease Prevention]. 2004. *Damp Indoor Spaces and Health.* Academy of Science. Washington, DC: The National Academies Press.

Jaakkola, J. J., B. F. Hwang, and N. Jaakkola. 2005. Home dampness and molds, parental atopy, and asthma in childhood: A six-year population-based cohort study. *Environ Health Perspect* 113(3):357–361.

Jarvis, B. B., W. G. Sorenson, and E. L. Hintikka 1998. Study of toxin production by isolates of Stachybotrys chartarum and Memnoniella echinata isolated during a study of pulmonary hemosiderosis in infants. *Appl Environ Microbiol* 64(10):3620–3625.

Kagen, S. L., J. N. Fink, D. P. Schlueter, et al. 1981. *Streptomyces albus*: A new cause of hypersensitivity pneumonitis. *J Allergy Clin Immunol* 68:295–299.

Karvonen, A. M., A. Hyvärinen, M. Korppi, et al. 2015. Moisture damage and asthma: A birth cohort study. *Pediatrics* 135(3):e598–606.

Koskinen, O., T. Husman, A. Hyvärinen, T. Reponen, and A. Nevalainen. 1995. Respiratory symptoms and infections among children in a day-care center with mold problems. *Indoor Air* 5(1):3–9.

Kurup, V. P., H. D. Shen, and H. Vijay. 2002. Immunobiology of fungal allergens. *Int Arch Allergy Immunol* 129(3):181–188.

Ławniczek-Wałczyk, A., and R. L. Górny. 2010. Endotoxins and β-glucans as markers of microbiological contamination: Characteristics, detection, and environmental exposure. *Ann Agric Environ Med* 17(2):193–208.

Lehtonen, M., T. Reponen, and A. Nevalainen. 1993. Everyday activities and variation of fungal spore concentrations in indoor air. *Int Biodeterior Biodegrad* 31(1):25–39.

Liebers, V., M. Raulf-Heimsoth, and T. Brüning. 2008. Health effects due to endotoxin inhalation (review). *Arch Toxicol* (Review) 82(4):203–210.

Mackiewicz, B. 2013. *Wpływ pyłów organicznych na układ oddechowy: Badania środowiskowe i kliniczne.* Lublin: Wydawnictwa Uniwersytetu Medycznego w Lublinie.

Meklin, T., T. Husman, A. Vepsäläinen et al. 2002. Indoor air microbes and respiratory symptoms of children in moisture damaged and reference schools. *Indoor Air* 12(3):175–183.

Mendell, M. J., W. J. Fisk, K. Kreiss et al. 2002. Improving the health of workers in indoor environments: Priority research needs for a national occupational research agenda. *Am J Public Health* 92(9):1430–1440.

Moses, L., K. Morrissey, R. A. Sharpe, and T. Taylor. 2019. Exposure to indoor mouldy odour increases the risk of asthma in older adults living in social housing. *Int J Environ Res Public Health* 16(14):2600.

Mudarri, D., and W. J. Fisk. 2007. Public health and economic impact of dampness and mold. *Indoor Air* 17(3):226–235.

Mudarri, D. H. 2016. Valuing the economic costs of allergic rhinitis, acute bronchitis, and asthma from exposure to indoor dampness and mold in the US. *J Environ Public Health* 2016:2386596.

Noe, R., A. L. Cohen, E. Lederman et al. 2007. Skin disorders among construction workers following Hurricane Katrina and Hurricane Rita: An outbreak investigation in New Orleans, Louisiana. *Arch Dermatol* 143(11):1393–1398.

Oluwole, O., S. P. Kirychuk, J. A. Lawson et al. 2016. Indoor mold levels and current asthma among school-aged children in Saskatchewan, Canada. *Indoor Air* 27(2):311–319.

Park, J., J. M. Cox-Ganser, S. K. White et al. 2017. Bacteria in a water-damaged building: Associations of actinomycetes and non-tuberculous mycobacteria with respiratory health in occupants. *Indoor Air* 27(1):24–33.

Platt, S. D., C. J. Martin, S. M. Hun, and C. W. Lewis. 1989. Damp housing, mold growth, and symptomatic health state. *Br Med J* 298(6689):1673–1678.

Quansah, R., M. S. Jaakkola, T. T. Hugg, S. A. M. Heikkinen, and J. J. K. Jaakkola. 2012. Residential dampness and molds and the risk of developing asthma: A systematic review and meta-analysis. *PLoS ONE* 7(11):e47526.

Rando, R. J., C. W. Kwon, and J. J. Lefante. 2014. Exposures to thoracic particulate matter, endotoxin, and glucan during post-Hurricane Katrina restoration work, New Orleans 2005–2012. *J Occup Environ Hyg* 11(1):9–18.

Rasmussen, M. K., P. Foldbjerg, and J. Christoffersen. 2017. Buildings and their impact on the health of Europeans. https://www.researchgate.net/publication/317256481_Healthy_Homes_Barometer_2017_-_Buildings_and_Their_Impact_on_the_Health_of_Europeans. [accessed October 11, 2019].

Rylander, R. 1994. Organic dusts-from knowledge to prevention. *Scand J Work Environ Health* 20(Spec No):116–122.

Rylander, R., and P. G. Holt. 1998. (1–3)-β-D-glucan and endotoxin modulate immune response to inhaled allergen. *Mediat Inflamm* 72:105–110.

Rylander, R., K. Persson, H. Goto, K. Yuasa, and S. Tanaka. 1992. Airborne beta-1,3-glucan may be related to symptoms in sick buildings. *Indoor Environ* 1(5):263–267.

Sauni, R., J. Uitti, M. Jauhiainen et al. 2013. Remediating buildings damaged by dampness and mould for preventing or reducing respiratory tract symptoms, infections and asthma (Review). *Evid Based Child Health* 8:3: 944–1000.

Shah, A., and C. Panjabi. 2014. Allergic aspergillosis of the respiratory tract. *Eur Respir Rev* 23(131):8–29.

Sharpe, R. A., N. Bearman, C. R. Thornton, K. Husk, and N. J. Osborne. 2015. Indoor fungal diversity and asthma: A meta-analysis and systematic review of risk factors. *J Allergy Clin Immunol* 135(1):110–122.

Sheehan, W. J., and W. Phipatanakul. 2016. Indoor allergen exposure and asthma outcomes. *Curr Opin Pediatr* 28(6):772–777.

Smedje, G., J. Wang, D. Norbäck, H. Nilsson, and K. Engvall. 2017. SBS symptoms in relation to dampness and ventilation in inspected single-family houses in Sweden. *Int Arch Occup Environ Health* 90(7):703–711.

Spengler, J. D., J. M. Samet, and J. F. McCarthy, eds. 2000. *Indoor Air Quality Handbook*. New York, NY: McGraw-Hill Book Co.

Thrasher, J. D., M. R. Gray, K. H. Kilburn, D. P. Dennis, and A. Yu. 2012. A water-damaged home and health of occupants: A case study. *J Environ Public Health* 2012:10.

Wang, J., Z. Zhao, Y. Zhang et al. 2019. Asthma, allergic rhinitis and eczema among parents of preschool children in relation to climate, and dampness and mold in dwellings in China. *Environ Int* 130:104910.

Yang, C. S., and E. Johanning. 1997. Airborne fungi and mycotoxins. In: *Manual of Environmental Microbiology*, ed. C. J. Hurst, 651–660. Washington, DC: ASM Press.

Zain, M. E. 2011. Impact of mycotoxins on humans and animals. *J Saudi Chem Soc* 15(2):129–144.

Zyska, B. 1999. *Zagrożenia biologiczne w budynku*. Warsaw: Arkady.

# 4 Environmental Conditions Affecting Microbiological Contamination of Buildings

*Małgorzata Gołofit-Szymczak*

## CONTENTS



The main sources of microorganisms on Earth are soil, plants, including agricultural crops and forests, wetlands, deserts, glaciers, urban areas, natural and anthropogenic reservoirs, and industry. There are significant differences in concentrations of microorganisms, depending on the environmental conditions, e.g. the average concentration of bacteria in urban areas is about $9.2 \times 10^5$ CFU/m$^3$, while in the desert it is about $3.8 \times 10^4$ CFU/m$^3$ [Burrows et al. 2009]. The atmospheric air transports a very large number of particles which belong to non-pathogenic saprophytic microbiota. These are mainly bacterial spores and fungal conidia. Microbiological particles enter the atmosphere as a result of being removed from the surface of plants and soil, because of wind pressure or thermal convection, after their spontaneous or forced rainfall emission from natural water reservoirs, and as a result of storage and processing of solid and liquid waste [Kulkarni et al. 2011; IOM 2004]. Their development and dissemination in the environment depends on their structure (e.g. shape, size etc.) and resulting functions (e.g. the ability to create spore forms), as well as on the various environmental parameters, associated with, inter alia, the geographical region (temperature, humidity), oxygen content, presence of organic and inorganic sources of nutrients, electromagnetic radiation, the flora and fauna, the population or socio-economic human activities (e.g. urbanisation of the area, agricultural crops, stock-breeding etc.) [Meadow et al. 2014; Macher 1999].

## 4.1 OUTDOOR ENVIRONMENTAL FACTORS

The outdoor environment has a large impact on microbial concentrations in buildings. The atmospheric air can penetrate premises in an uncontrolled manner as a result of a leakage in the building envelope and in a controlled manner through ventilation systems. There are several factors underpinning the growth, development and spread of microorganisms. Some of the most important include season (meteorological conditions: temperature, precipitation, wind speed) and geographical location. The season has a significant impact on the qualitative and quantitative composition of indoor air mycobiota (the influence on the bacteria is not so noticeable) [Adams et al. 2013; Pitkaranta et al. 2008]. In the scientific literature, there are quite a number of studies characterising changes in bioaerosol concentrations depending on the time of year (e.g. Bernasconi et al. 2010; Ejdys and Biedunkiewicz 2011; Giulio et al. 2010; Grinn-Gofroń 2011; Kummer and Thiel 2008; Jo and Seo 2005; Reponen et al. 1992; Tseng et al. 2011). Their analysis indicates that atmospheric concentrations of bacteria in the summer season range from $10^2$ to $10^4$ CFU/m$^3$, while in winter they are lower and usually do not exceed $10^2$ CFU/m$^3$. In the indoor environment, bacterial aerosol concentrations range from $10^2$ to $10^5$ CFU/m$^3$ in the summer and from $10^2$ to $10^3$ CFU/m$^3$ in the winter. Fungal aerosol, similarly to bacterial aerosol, also shows quantitative differences in different seasons. Fungal concentrations in the atmospheric air in summer vary from $10^2$ to $10^4$ CFU/m$^3$, and in winter they are below $10^2$ CFU/m$^3$. However, in the indoor environment, concentration levels of this type of aerosol usually do not exceed $10^4$ CFU/m$^3$ in summer and $10^3$ CFU/m$^3$ in winter. In summer, higher concentrations of bacterial and fungal aerosols are observed in the temperate zone, both in indoor and outdoor air. Many authors suggest that high temperatures with strong UV radiation in summer can also lower the level of microorganisms in the air [Chi and Li 2007; Wu et al. 2012]. Sunlight has a bactericidal effect due to the portion of the range of radiation that includes the short-wave violet and ultraviolet radiation [Cox et al. 1995]. Meteorological conditions, such as wind speed and direction, can also have a significant impact on microbial concentrations in the air. Strong bursts of wind can release microorganisms from the ground and plants and cause them to be spread widely [Jones and Harrison 2004].

The studies of Amend et al. [2010] indicated that the location of the building, depending on the latitude, also has a significant influence on the qualitative composition of mycobiota present indoors, and this in turn correlates with local atmospheric conditions such as precipitation and the availability of nutriments in the form of organic matter. In the countries in subtropical and tropical zones, during the rainy season, higher levels of saprophytic mould (*Aspergillus*, *Penicillium*, *Chaetomium* and *Cladosporium*) concentrations can be observed.

Another element that has a fundamental impact on the microbiological quality of the outdoor air and thus influences the indoor environment, is urbanisation. A study by Shaffer and Lighthart [1997] showed that the average bacterial concentrations in urban and rural areas of Oregon, in the US, were 609 CFU/m$^3$ versus 242 CFU/m$^3$.

In France, the average bacterial level in Marseille was 18 times higher than in the nature reserve on the island of Porquerolles (791 CFU/m$^3$ versus 42 CFU/m$^3$) [Di Giorgio et al. 1996].

Municipal waste landfills, sewage treatment plants or biomass processing in urban areas are active sources of organic dust containing numerous harmful microbial agents. In the case of landfills, bioaerosol emission can take place at different stages of waste management, both during transport and unloading, as well as during sorting and storage. These activities may cause an increase in the concentration of microorganisms, especially when a given landfill is above ground level and is not surrounded by trees [Skibniewska 2011].

## 4.2 MICROCLIMATE OF THE INDOOR ENVIRONMENT

The key environmental parameters that determine the development and survival of microorganisms are air temperature and relative humidity (RH). Temperature strictly determines water activity and, as a physical parameter of the environment, can directly or indirectly affect microbial growth. Microorganisms do not have internal temperature control mechanisms and within the cell the temperature is determined by the conditions outside it. The majority of the bacteria widespread in nature belong to the group of mesophilic organisms, whose optimum growth varies, depending on the species, between 20–40°C [Macher 1999]. Moulds and actinomycetes have a high temperature tolerance. For the most common indoor moulds, the maximum growth is within the range of 22–35°C, but they can also grow in low (from 5°C to 10°C) and high (from 35°C to 52°C) temperatures. In temperatures from 50°C to 60°C, the growth of most moulds slows down and stops entirely. In the case of actinomycetes, the optimal growth is between 22–35°C or exceeds this range significantly (e.g. in case of some thermophilic species like *Thermoactinomyces*, the optimal growth temperature is between 50–60°C) [Holt et al. 1994].

## 4.3 HUMIDITY AS A KEY INITIATOR OF MICROBIOLOGICAL CONTAMINATION

Indoor air humidity usually changes with the humidity in the outdoor environment, although it is also affected by numerous indoor sources such as sanitary and kitchen fittings, aquariums, ventilation and heating systems. The emission from living organisms (humans, animals, plants) plays also certain role here [Adan and Samson 2011]. The structure of a building is essentially intended to keep it relatively dry; however, its failure can result in a significant humidification of the indoor spaces. Water can enter buildings through cracks or leaks in the plumbing system, leaks in the roof and other structural elements of the building envelope or can be the result of steam condensation on cold surfaces (poorly insulated walls or floors) or can enter the building as water pulled up from the soil by capillary forces [Prezant et al. 2008] (Figure 4.1). Increased humidity in buildings can also have external sources deriving from floods, storms, hurricanes and typhoons or heavy rainfalls. Indoor humidity may result in a massive microbial growth, which in turn can have an adverse effect on human health [Barbeau et al. 2010; Bloom et al. 2009; Chew et al. 2006]. Numerous studies

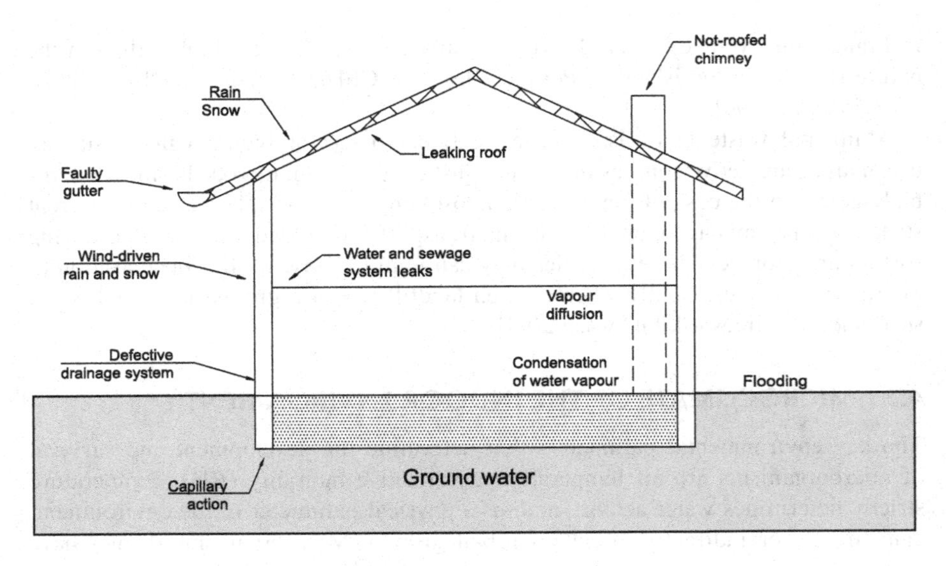


**FIGURE 4.1** Main sources of building dampness.

show a positive correlation between air humidity and the number of microorganisms indoors (e.g. Holmquist et al. 1983; Dubey and Jain 2014).

The kind of microorganism that will colonise particular spaces or surfaces in the indoor environment depends on the physical and chemical characteristics of the construction and finishing materials used in a given building and the nutrients that the individual components of these materials can become for the microorganisms, but above all it depends on the extent to which a given material is able to meet the requirements of the specific microorganism in terms of the amount of moisture necessary to initiate its growth and maintain its subsequent development. In microbiology, this value is described by the so-called 'water activity' ($a_w$), also known as equilibrium relative humidity (ERH), which is the proportion of water vapour pressure in a certain material to clean water vapour pressure at the same temperature and pressure [Macher 1999]. As shown by numerous studies, an $a_w$ rate of 0.65 (which equates to ERH = 65%) is the lowest rate necessary for initiating the microbial growth on a certain material which contains enough nutrients. The absolute lower limit for $a_w$ is 0.55, the point at which deoxyribonucleic acid (DNA) is denatured [Griffin 1981]. The activity of microorganisms and their ability to colonise new surfaces increases when $a_w$ approaches 1, i.e. when water is freely available [Lacey and Dutkiewicz 1994]. Following the water activity parameter, the microorganisms can be categorised according to ability to initiate growth on building materials and arranged in the order in which they can appear on the surfaces (as primary, secondary and tertiary colonisers) [Adan et al. 1994]. Although the high humidity levels, as well as the surface and intra-structural condensation, are sufficient for primary and secondary colonisers, organisms in the third group require a significant presence of water in the colonised environment. Such availability of water usually occurs in the event of water failure resulting from structural defects of the building or water network, improper insulation, water intrusion or flooding. The maximum tolerance of extreme temperatures is visible near the optimal rates of water activity. Based on the rates of the $a_w$ parameter, microorganisms can be classified as primary ($a_w$ <0.85, ERH <85%), secondary ($a_w$ = 0.85–0.90, ERH = 85–90%) and tertiary ($a_w$ >0.90, ERH >90%) colonisers [Górny 2004b; Grant et al. 1989]. Primary (xerophilic) colonisers include the following fungi: *Alternaria citri*, numerous species of the *Aspergillus* genus (*A. candidus, A. niger, A. ochraceus, A. penicillioides, A. restrictus, A. rubrobrunneus, A. sydowii, A. terreus, A. wentii*), *Eurotium echinulatum, Paecilomyces variotii, Penicillium* species (*P. aurantiogriseum, P. brevicompactum, P. chrysogenum, P. citrinum, P. expansum*) and *Wallemia sebi*. Secondary colonisers include the following fungi: *Absidia corymbifera, Aspergillus* species (*A. clavatus, A. flavus, A. versicolor*), *Aureobasidium pullulans, Chrysonilia sitophila*, species of *Cladosporium* genus (*C. herbarum, C. cladosporioides, C. sphaerospermum*), *Epicoccum nigrum*, species of *Fusarium* genus (*F. culmorum, F. graminearum, F. solani*), *Mucor circinelloides, Penicillium oxalicum, Rhizopus oryzae, Ulocladium chartarum* and *Verticillium lecanii*. Tertiary (hydrophilic) colonisers include the following microorganisms: *Alternaria alternata, Aspergillus fumigatus, Botrytis cinerea, Epicoccum* spp., *Exophiala* spp., *Fusarium moniliforme, Geomyces pannorum, Mucor plumbeus, Mucor racemosus, Neosartorya fischeri, Phoma herbarum, Phialophora* spp., *Rhizopus stolonifera, Rhodotorula* spp., *Sistotrema brinkmannii, Stachybotrys chartarum, Trichoderma* spp., *Ulocladium consortiale, Sporobolomyces* spp. and actinomycetes.

## 4.4 VENTILATION OF BUILDINGS

Today, it is quite common in the world to improve indoor air quality using various technical solutions. Ventilation systems are one of them. Modern ventilation systems, among other numerous pro-health applications, allow reduction in microbiological air contamination. Users of ventilated rooms should be certain that the air provided to them is of a good quality. At the same time, they should be informed about possible health risks resulting from inhalation of polluted air [WHO 2000]. Ventilation systems should ensure an adequate air quality in the whole building, in a specific room or in a part of it. The most important air parameters that can be regulated in a ventilated room are temperature, relative humidity and concentration of pollutants. The rates of these parameters depend on the system used for air conditioning. Figure 4.2 shows the classification of ventilation systems and their structures.

Ventilation systems should deliver clean air and by that provide dilution and/or removal of the pollutants present indoors. However, over time, they may become contaminated and as such, become themselves a source of microbiological air contamination. Bacteria and moulds, as relatively small aerodynamic structures, can enter ventilated premises in this way. In practice, due to excellent colonisation abilities, those microorganisms may turn every element of the system (ventilation ducts, air filters, thermal insulation, noise silencers or air coolers) into the element supporting their growth and survival.

Ventilation systems installed in buildings have a significant influence on microbial indoor air quality. The studies performed by Gołofit-Szymczak and Górny

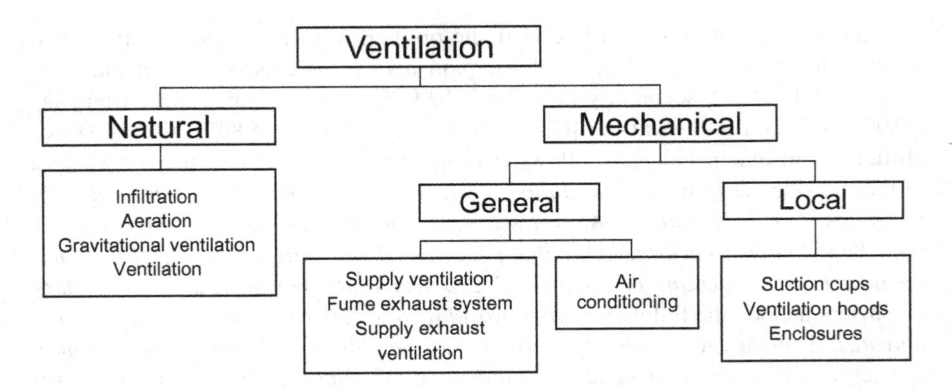


**FIGURE 4.2** Classification of ventilation systems and their structures.

[2018] showed that the highest bioaerosol concentrations were recorded in premises with natural/gravitational ventilation reaching $1.6 \times 10^3$ CFU/m$^3$. In turn, in the rooms with mechanical ventilation or central air-conditioning systems (e.g. in office buildings), the concentrations of microorganisms in the air were lower and did not exceed $5.3 \times 10^2$ CFU/m$^3$.

### 4.4.1 Natural Ventilation

Natural ventilation stirs the air masses within the building using natural forces (wind, gravitation, temperature differences) and construction features of ventilated premises. Natural ventilation includes [Allard and Santamouris 1998; Yamanaka et al. 2006]:

a) infiltration – air penetration through porous building materials, leaks in partitions and gaps around windows and doors;
b) aeration – organised exchange of the air, which enters rooms through openings located in external building partitions as a result of the difference in indoor and outdoor air pressure;
c) gravity ventilation – air exchange through vertical exhaust ducts as a result of the difference in indoor and outdoor air density;
d) airing – periodic exchange of the air in rooms by opening windows, doors or other openings in building partitions.

### 4.4.2 Mechanical Ventilation

In this type of ventilation, the air exchange is independent of any weather conditions. Forced air flow is achieved by using a mechanical means of stimulating air movement in the form of a fan or ejector (jet pump). Due to mechanical ventilation, it is possible to effectively influence the room temperature, achieve the assumed pressure distribution, control the speed of air movement and remove larger amounts of pollutants. Mechanical ventilation can have many technical variations depending on: the way the air is exchanged, the direction of air movement in relation to the ventilated interior and the indoor/outdoor pressure difference.

The ventilation system consists of supply and exhaust parts. The supply unit includes air intake, air filter, heater, fans, supply-air outlets and ducts. The exhaust air unit consists of exhaust air vents, ducts, fans and air outlets. Depending on the direction of the air in relation to the ventilated room, one can distinguish [Awbi 2007; Cao et al. 2014]:

a) supply ventilation – the air supply is mechanical, and its removal is natural through the leaks in windows and doors or using ventilators;
b) exhaust ventilation – the air is supplied in a natural way and only the exhaust is mechanically supported;
c) supply and exhaust ventilation – both elements directing the air are fully mechanical.

### 4.4.3 Air-Conditioning System

Air-conditioning is the process of adding to the air certain parameters and properties desired for hygienic reasons, taking into account the well-being of exposed individuals ('comfort air-conditioning') or the rates of parameters required by production technologies ('industrial air-conditioning'). Air-conditioning is the best form of mechanical ventilation system, equipped with a full set of versatile devices, which helps to maintain the required level of crucial parameters (temperature, humidity, cleanliness and air movement within the ventilated space) throughout the year, regardless of changes and fluctuations of air parameters occurring outside the building.

Air-conditioning systems include ventilation devices, spray chambers (cooling and drying the air) and an extensive system for automatic regulation of air temperature and humidity [Awbi 2007; Cao et al. 2014]. The external air is drawn into the buildings equipped with a ventilation system and is conditioned in order to eliminate the biological and chemical pollutants present in it and to ensure optimal temperature and humidity. One of the most common methods of conditioning is filtration of the air on non-woven filters before it is brought into the rooms. The aerosol particles carried with the outdoor air are partially retained by the air filters placed in the system inlet. The efficiency of air purification using such pre-filters is limited. Their main task is to protect the conditioning equipment from larger pollutants present in the outdoor air. Particles that have not been retained on the pre-filters may settle on the surfaces in air handling units, in ducts or on other elements of ventilation systems causing their contamination [Brosseau et al. 2000; Chang et al. 1996; Charkowska 2003; Zuraimi 2010].

Improperly maintained ventilation systems (e.g. due to the use of filters with too low particle capture efficiency, long-term operation of filters or lack of regular cleaning or disinfection) may cause additional contamination of rooms as a result of secondary dusting [Gołofit-Szymczak and Górny 2018]. Some of the pollutants deposited on the internal surfaces of ventilation ducts may be redistributed on other elements of the installation causing their secondary contamination.

**TABLE 4.1**
**Selected Elements of Ventilation Systems and Microorganisms Responsible for Their Colonisation**

| Components of the ventilation system | Microbial genera |
|---|---|
| Duct surfaces | *Aspergillus, Cladosporium, Trichoderma, Penicillium, Rhizopus, Rhodotrula, Bacillus, Micrococcus, Staphylococcus* |
| Air coolers | *Aspergillus, Cladosporium, Penicillium* |
| Droplet separators, dehydrators, siphons | *Legionella, Thermoactinomycetes, Penicillium, Alternaria, Acremonium* |
| Filters | *Penicillium, Aspergillus, Acremonium, Cladosporium, Bacillus, Micrococcus, Staphylococcus, Pseudomonas* |
| Cooling towers | *Legionella, Thermoactinomycetes, Acremonium, Penicillium, Cladosporium, Aspergillus* |

In the scientific literature, there are numerous studies characterising microbiological contamination of ventilation systems. Their analysis shows that among bacteria, the most frequently isolated are species of the genus: *Bacillus* (*B. cereus, B. pumilus*), *Micrococcus* (*M. luteus, M. roseus*), *Staphylococus* (S. *epidermidis, S. saprophyticus, S. hominis, S. capitis*), *Pseudomonas, Flavobacterium, Acinetobacter, Alcaligenes*, meso- and thermophilic actinomycetes, and among fungi, the species of the genus: *Penicillium* (*P. expansum, P. chrysogenum, P. commune, P. citrinum*), *Aspergillus* (*A. fumigatus, A. flavus*), *Acremonium, Cladosporium, Alterneria* (*A. alternata, A. tenuissiuma*) and *Fusarium* (*F. proliferatum*) [Brickus et al. 1998; Gołofit-Szymczak and Górny 2010; Kanaani et al. 2008; Noris et al. 2011; Pasanen et al. 1993; Zuraimi 2010]. Table 4.1 shows selected elements of ventilation systems and the microorganisms that usually inhabit them [Charkowska 2003].

Water used in air cooling devices of the air-conditioning systems may over time become a perfect medium supporting fungal growth. In turn, ventilation system components that generate water aerosol (e.g. cooling towers, air coolers, evaporator exchangers, spray chambers, sprinkler plate exchangers) also create favourable conditions for the development of bacteria, including pathogenic *Legionella* strains [Brickus et al. 1998; Charkowska 2003; Price et al. 2005].

## REFERENCES

Adams, R., M. Miletto, J. W. Taylor, and T. D. Bruns. 2013. Dispersal in microbes: Fungi in indoor air are dominated by outdoor air and show dispersal limitation at short distances. *ISME J* 7(7):1262–1273.

Adan, O. C. G., R. A. Samson, and J. T. M. Wijnen 1994. Fungal resistance tests: A proposed method for testing resistance of interior finishes Air quality monographs. In: *2: Health Implications of Fungi in Indoor Environments*, eds. R. A. Samson, B. Flannigan, M. E. Flannigan, A. P. Verhoeff, O. C. G. Adan, and E. S. Hoekstra, 415–437. Amsterdam: Elsevier Science B.V.

Adan, O. C. G., and R. A. Samson. 2011. *Fundamentals of Mold Growth in Indoor Environments and Strategies for Healthy Living.* The Netherlands: Wageningen Academic Publishers.

Allard, F., and. M. Santamouris, eds. 1998. *Natural Ventilation in Buildings: A Design Handbook.* 1st ed. London: Earthscan Publications, Ltd.

Amend, A. S., K. A. Seifert., R. Samson, and T. D. Bruns. 2010. Indoor fungal composition is geographically patterned and more diverse in temperate zones than in the tropics. *Proceedings of the National Academy of Sciences of the United States of America, 107*(31): 13748–13753.

Awbi, H. B. 2007. *Ventilation Systems: Design and Performance.* London: Spon Press.

Barbeau, D. N., L. F. Grimsley, L. E. White, J. M. El-Dahr, and M. Lichtveld. 2010. Mold exposure and health effects following hurricanes Katrina and Rita. *Annu Rev Public Health* 31:165–179.

Bernasconi, C., M. Rodolfi, A. M. Picco, P. Grisoli, C. Dacarro, and D. Rembges. 2010. Pyrogenic activity of air to characterise bioaerosol exposure in public buildings: A pilot study. *Lett Appl Microbiol* 50(6):571–577.

Bloom, E., L. F. Grimsley, C. Pehrson, J. Lewis, and L. Larsson. 2009. Molds and mycotoxins in dust from water-damaged homes in New Orleans after Hurricane Katrina. *Indoor Air* 19(2):153–158.

Brickus, L. S. R., L. F. G. Siqueira, F. R. Aquino Neto, and J. N. Cardoso. 1998. Occurrence of airborne bacteria and fungi in bayside offices in Rio de Janeiro, Brazil. *Indoor Built Environ* 7(5–6):270–275.

Brosseau, L. M., D. Vesley, T. H. Kuehn, J. Melson, and H. S. Han. 2000. Methods and criteria for cleaning contaminated ducts and air-handling equipment. *ASHRAE Trans* 106:188–189.

Burrows, S. M., W. Elbert, M. G. Lawrence, and U. Pöschl. 2009. Bacteria in the global atmosphere, Part 1: Review and synthesis of literature data for different ecosystems. *Atmos Chem Phys Discuss* 9(23):9263–9280.

Cao, G., H. Awbi, R. Yao et al. 2014. A review of the performance of different ventilation and airflow distribution systems in buildings. *Build Environ* 73:171–186.

Chang, J. C. S., K. K. Foarde, and D. W. VanOsdell. 1996. Assessment of fungal (Penicillium chrysogenum) growth oh three HVAC duct materials. *Environ Int* 22(4):425–431.

Charkowska, A. 2003. *Zanieczyszczenia w instalacjach klimatyzacyjnych i metody ich usuwania.* Gdańsk: IPPU MASTA.

Chew, G. L., J. Wilson, F. A. Rabito et al. 2006. Mold and endotoxin levels in the aftermath of Hurricane Katrina: A pilot project of homes in New Orleans undergoing renovation. *Environ Health Perspect* 114(12):1883–1889.

Chi, M. C., and C. S. Li. 2007. Fluorochrome in monitoring atmospheric bioaerosols and correlations with meteorological factors and air pollutants. *Aerosol Sci Tech* 41(7):672–678.

Cox, C. S., and C. M. Wathes, eds. 1995. *Bioaerosols Handbook.* Boca Raton, FL: Lewis Publishers/CRC Press, Inc.

Di Giorgio, C., A. Krempff, H. Guiraud, P. Binder, C. Tiret, and G. Dumenil. 1996. Atmospheric pollution by airborne microorganisms in the city of Marseilles. *Atmos Environ* 30(1):155–160.

Dubey, S., and K. Jain. 2014. Effect of humidity on fungal deteriogens of ancient monuments. *Int Res J Biol Sci* 3(4):84–86.

Ejdys, E., and A. Biedunkiewicz. 2011. Fungi of the genus Penicillium in school building. *Pol J Environ Stud* 20:333–338.

Giulio, M., R. Grande, E. Campli, S. Bartolomeo, and L. Cellini. 2010. Indoor air quality in university environments. *Environ Monit Assess* 170(1–4):509–517.

Gołofit-Szymczak, M., and R. L. Górny. 2010. Bacterial and fungal aerosols in air-conditioned office buildings in Warsaw, Poland, the winter season. *Int J Occup Saf Ergon* 16(4):465–476.

Gołofit-Szymczak, M., and R. L. Górny. 2018. Microbiological air quality in office buildings equipped with ventilation systems. *Indoor Air* 28(6):792–805.

Górny, R. L. 2004b. *Cząstki grzybów i bakterii jako składniki aerozolu pomieszczeń: Właściwości, mechanizmy emisji, detekcja.* Sosnowiec: Wyd. Instytutu Medycyny Pracy i Zdrowia Środowiskowego.

Grant, C., C. A. Hunter, B. Flannigan, and A. F. Bravery. 1989. The moisturise requirements of moulds isolated from domestic dwellings. *Int Biodeterior* 25(4):259–289.

Griffin, D. H. 1981. *Fungal Physiology.* New York: Wiley.

Grinn-Gofroń, A. 2011. Airborne Aspergillus and Penicillium in atmosphere of Szczecin, (Poland) (2004–2009). *Aerobiologia* 27(1):67–76.

Holmquist, G. U., H. W. Walker, and H. M. Stahr. 1983. Influence of temperature, pH, water activity and antifungal agents on growth of Aspergillus flavus and A. parasiticus. *J Food Sci* 48(3):778–782.

Holt, J. G., N. R. Krieg, P. H. A. Sneath, J. T. Stanley, and S. T. Williams, eds. 1994. *Bergey's Manual of Determinative Bacteriology.* Baltimore: Williams and Wilkins.

IOM [Institute of Medicine] 2004. *Damp Indoor Spaces and Health.* Washington, DC: The National Academies Press.

Jo, W. K., and Y. J. Seo. 2005. Indoor and outdoor bioaerosol levels at recreation facilities, elementary schools, and homes. *Chemosphere* 61(11):1570–1579.

Jones, A. M., and R. M. Harrison. 2004. The effects of meteorological factors on atmospheric bioaerosol concentrations – A review. *Sci Total Environ* 326(1–3):151–180.

Kanaani, H., M. Hargreaves, Z. Ristovski, and L. Morawska. 2008. Deposition rates of fungal spores in indoor environments, factors affecting them and comparison with non-biological aerosols. *Atmos Environ* 42(30):7141–7154.

Kulkarni, P., P. A. Baron, and K. Willeke, eds. 2011. *Aerosol Measurement: Principles, Techniques, and Applications.* New York: John Wiley and Sons, Inc.

Kummer, V., and W. R. Thiel. 2008. Bioaerosols: Sources and control measures. *Int J Hyg Environ Health* 211(3–4):299–307.

Lacey, J., and J. Dutkiewicz. 1994. Bioaerosols and occupational lung disease. *J Aerosol Sci* 25(8):1371–1404.

Macher, J., eds. 1999. *Bioaerosols: Assessment and Control.* American Conference of Governmental Industrial Hygienists, Cincinnati.

Meadow, J. F., A. E. Altrichter, S. W. Kembel et al. 2014. Indoor airborne bacterial communities are influenced by ventilation, occupancy, and outdoor air source. *Indoor Air* 24(1):41–48.

Noris, F., J. A. Siegel, and K. A. Kinney. 2011. Evaluation of HVAC filters as a sampling mechanism for indoor microbial communities. *Atmos Environ* 45(2):338–346.

Pasanen, P., A. L. Pasanen, and M. J. Jantunen. 1993. Water condensation promotes fungal growth in ventilation ducts. *Indoor Air* 3(2):106–112.

Pitkaranta, M., T. Meklin, A. Hyvarinen et al. 2008. Analysis of fungal flora in indoor dust by ribosomal DNA sequence analysis, quantitative PCR, and culture. *Appl Environ Microbiol* 74(1):233–244.

Prezant, B., D. Weekes, and J. D. Miller. 2008. *Recognition, Evaluation and Control of Indoor Mold. American Industrial Hygiene Association.* Fairfax, VA: American Industral Hygiene Assocation.

Price, D. L., R. B. Simmons, S. A. Crow, and D. G. Ahearn. 2005. Mold colonization during use of preservative-treated and untreated air filters, including HEPA filters from hospitals and commercial locations over an 8-year period (1996–2003). *J Ind Microbiol Biotechnol* 32(7):319–321.

Reponen, T., A. Nevalainen, M. Jantunen, M. Pellikka, and P. Kalliokoski. 1992. Normal range criteria for indoor air bacteria and fungal spores in subarctic climate. *Indoor Air* 2(1):26–31.

Shaffer, B. T., and B. Lighthart. 1997. Survey of culturable airborne bacteria at four diverse locations in Oregon: Urban, rural, forest, and coastal. *Microb Ecol* 34(3):167–177.

Skibniewska, K., ed. 2011. *Some Aspects of Environmental Impact of Waste Dumps*. Olsztyn: Department of Land Reclamation and Environmental Management. University of Warmia and Mazury.

Tseng, C. H., H. C. Wang, N. Y. Xiao, and Y. M. Chang. 2011. Examining the feasibility of prediction models by monitoring data and management data for bioaerosols inside office buildings. *Build Environ* 46(12):2578–2589.

WHO [World Health Organization]. 2000. *The Right to Healthy Indoor Air: Report on a WHO Meeting, Bilthoven, The Netherlands 15–17 May 2000*. Copenhagen: WHO Regional Office for Europe. https://apps.who.int/iris/handle/10665/108327. [accessed October 11, 2019].

Wu, Y. H., C. C. Chan, G. L. Chew, P. W. Shih, C. T. Lee, and H. J. Chao. 2012. Meteorological factors and ambient bacterial levels in a subtropical urban environment. *Int J Biometeorol* 56(6):1001–1009.

Yamanaka, T., H. Kotani, K. Iwamoto, and M. Kato. 2006. Natural, wind-forced ventilation caused by turbulence in a room with a single opening. *Int J Vent* 5(1):179–187.

Zuraimi, M. S. 2010. Is ventilation duct clearing useful? A review of scientific evidence. *Indoor Air* 20(6):445–457.

# 5 Biodeterioration of Building Materials

*Marcin Cyprowski*

## CONTENTS



## 5.1 BIODETERIORATION OF THE SURFACE

Moulds are extremely flexible in terms of their nutritional requirements and have great adaptability. They obtain basic nutrients (rich in carbon and nitrogen) by decomposition of organic matter [Macher 1999]. Most of the fungi present in the indoor environment are saprophytes, which mean that they extract nutrients from damp materials such as wood, paper, paints, adhesives, soil, dust, food particles, plant debris etc. However, they can grow with equal success on surfaces composed of inorganic matter (such as glass, fibreglass, metal or concrete) which are damp, covered with dust or even fingerprints that form a thin, invisible layer of biofilm [Sedlbauer 2001; Kołwzan 2011]. This is also the case with actinomycetes. These bacteria play an important role in the decomposition of many organic compounds, including lignin, cellulose, pectin, chitin, keratin, collagen, elastin and starch [Karbowska-Berent and Strzelczyk 2000].

During the growth of their colonies, both above-mentioned groups of microorganisms produce and release many powerful enzymes and acids which can very efficiently bring the matter that serves as a substrate to complete decomposition or partial disintegration. The fungi include highly cellulolytic (e.g. *Trichoderma*, *Botrytis*, *Chaetomium*, *Alternaria*, *Stemphylium*), proteolytic (e.g. *Mucor*, *Chaetomium*, *Aureobasidium*, *Gymnoascus*, *Trichoderma*, *Verticillium* and *Epicoccum*) as well as lipolytic microorganisms (as above and *Paecilomyces*) [Florian 2000; Gravesen et al. 1994; Karunasena et al. 2000; Szczepanowska and Lovett 1992]. Under certain conditions, moulds can also produce mycotoxins [Tuomi et al. 2000a, 2000b]. In relation to actinomycetes (especially those of *Streptomyces* genus), their proteo- and collagenolytic abilities are emphasised [Karbowska-Berent and Strzelczyk 2000; Peczyńska-Czoch and Mordarski 1988].

Man-made building and finishing materials used in construction are not an ideal source of nutrients to support microbial growth. However, this does not mean that they prevent such growth. In nature, there are many habitats poor in such nutrients. Their absence is a factor rather for enforcing adaptive selection among the microorganisms than for slowing down their growth [Gooday 1988]. Analysing the situation in this regard, buildings are only one of the possible habitats, creating specific conditions, in which many microorganisms can successfully find their niche, and the very process of biodeterioration, being initiated by individual microorganisms, can over time lead to the formation of a complex ecosystem [Burge 1995].

Usually only materials rich in carbon can ensure microbial growth at sufficiently high humidity values [Nielsen et al. 2000]. When the growth is to take place on inorganic materials such as mineral wool or concrete, water intrusion, which would carry some organic matter, is necessary. In practice, low-density materials (<200 kg/m$^3$) have a more porous structure, which creates a much larger surface area on which the deposition and subsequent microbial growth can take place. High-density materials (>1000 kg/m$^3$) are less sensitive to moisture absorption [Reiman et al. 2000]. In the scientific literature, there are quite numerous reports showing the colonization possibilities of microorganisms in relation to construction and finishing building materials. Moulds and actinomycetes can develop on plaster, brick and concrete walls, damp wood, wood-based materials (particleboard), plasterboards, ceiling panels, wallpapers, paints, adhesives, floorings, carpets etc.

## 5.2 BIODETERIORATION OF CONSTRUCTION AND FINISHING MATERIAL STRUCTURE

Microbiological corrosion of materials constitutes a problem when it comes to the durability of a building, its aesthetics and the health of people living in it. It is a multi-stage and complex process of destroying materials used in construction by viable organisms, mainly fungi and bacteria. This phenomenon is often called biodeterioration, which means a decrease in the quality of building materials as a consequence of biological agents [Wołejko and Matejczyk 2011].

Fungi, especially moulds, are considered to be the main factor in biodeterioration processes. They attack mainly wood of both coniferous and deciduous species. They also cause bio-corrosion of other organic materials, such as fibreboards, particleboards, flaxboards, reed and straw mats, floorings, wallpapers, distemper etc. [e.g. Becker 1984; Ciferri 1999; Ellringer et al. 2000; Hyvärinen et al. 2002; Kujanpää et al. 1999; Pessi et al. 2002; Zyska 1999]. Bacteria, on the other hand, are often pioneering microorganisms in the biodeterioration of external building components such as brick, concrete, mortar, stones etc. [Piontek and Lechów 2013; Zyska 1999]. The available data indicates that the intensity of colonisation of building materials by various types of microorganisms may be largely due to the geographical location of the buildings and the porosity of materials, as well as specific areas in the body of the building, such as facade niches or sharp bends [Gaylarde and Gaylarde 2005]. However, regardless of the course of these processes, they ultimately lead to changes in indoor air quality that are dangerous to human health or even life.

## 5.3 SIGNS OF BIO-CORROSION

Depending on the type of material and conditions for development, microbial corrosion is accompanied by signs that can be divided into two main groups: morphological changes and changes in material properties. Morphological signs are those which can be subjectively assessed by the user of a given building or premise. These include:

- growth of fungi or actinomycetes, which is of varying intensity,
- discolouration of materials,
- deformation and peeling of paint coatings,
- decomposition/softening of the structure of paper wallpapers,
- fibrillation/loosening of the structure of paper, wood and wood-based materials.

When carrying out such inspections, particular attention should be paid to the specific locations of bio-corrosion in buildings. Moulds are usually located in the corners of external and partition walls, ceiling partitions, basements, on the ground floor and the last floor of the building, on the walls next to window and door frames, in the lower parts of the partition walls adjacent to staircases and ceilings in lavatories.

Signs of changes in material properties can, to some extent, be assessed by the building users themselves. However, such assessment often requires a number of laboratory analyses to examine chemical (e.g. the degree of polymerisation of the paper), mechanical (e.g. decrease in the strength of wood), electrical (e.g. decrease in resistivity of electrical insulating materials) or optical (e.g. opalescent glass) properties of building materials. This group of signs also includes changes in the organoleptic properties of materials, including e.g. the emission of odours associated with an increase in their moisture content and microbial decomposition [Zyska 1999].

## 5.4 BUILDING MATERIALS SUBJECT TO MICROBIAL CORROSION

It is not possible to investigate the causes of bio-corrosion of building materials without knowing the causative agent. For this purpose, it is necessary to trace the process of microbial decomposition, in which three phases are distinguished: infection (period from the germination of the microorganism until establishing permanent contact with the material), incubation (period from the end of infection to the occurrence of the first signs of material decomposition) and decomposition (period from the occurrence of the first signs to the complete decomposition of materials) [Zyska 1999]. Below is a brief description of the most common building materials and biological agents responsible for their bio-corrosion.

1) Wood: The main causes of bio-corrosion are fungi, which can develop even at wood moisture content of approximately 30%. Based on the way they work, they can be divided into two groups: fungi causing the rotting of wood, and fungi causing its permanent colour changes. Fungi of the first group have the ability to break down the cellulose contained in the wood, which contributes to significant weakening and/or permanent damage to

the wood elements. Their most common representative in Europe is *Serpula lacrymans*, which decomposes soft, coniferous wood. The second group consists of fungi that can penetrate the structure of wooden elements, permanently changing their natural colour (e.g. to blue), contributing significantly to lowering its quality. This group includes moulds (mainly those of *Fusarium* and *Penicillium* genera), which most often develop on the surface of the above-mentioned elements. These moulds do not reduce the strength of the wood directly, but the colonies densely occurring on the surface increase water absorption, and can therefore create ideal conditions for the growth of fungi causing rotting processes [Schmidt and Kallow 2005; Schmidt 2007; Clausen and Yang 2007].

2) Stone: Its corrosion is a lengthy process (usually lasting decades), initiated by microorganisms, lichens and mosses. The porosity of the structure of the material (which can reach 18%) is a factor that favours colonisation. When the pores of the material are sufficiently damp, they are penetrated by bacteria, which in their metabolism use inorganic sources of hydrogen in the form of ammonia, nitrogen dioxide (e.g. *Nitrosomonas* and *Nitrobacter* genera, which produce nitrous and nitric acid, respectively) as well as hydrogen sulphide and elemental sulphur (*Thiobacillus* genus, which produces sulphuric acid) [Saiz-Jimenez 1994]. Apart from bacteria, fungi, which contribute to the chemical and mechanical decomposition of stone, play an important role here. The most frequently mentioned in this context are moulds of *Penicillium*, *Cladosporium*, *Fusarium*, *Phoma* and *Trichoderma* genera [Dakal and Cameotra 2012]. Algae, lichens and plants also participate in the bio-corrosion processes [Saiz-Jimenez 1994]. Biodeterioration causes the loss of stone cohesion, an increase in its porosity, discolouration and efflorescence on its surface. All this leads to changes in the hygrothermal properties, which destroy this material.
3) Bricks and mortars: Products with a porous structure are most often exposed to biodeterioration, especially low-quality bricks with a moisture content of up to 3%. Lime mortars are exposed to nitrifying bacteria more often than cement mortars. The carbonate content and sulphur compounds reacting with the products of bacterial metabolic activity play an important role in the intensification of corrosion processes. Hence, the bio-corrosion of these materials is often caused by sulphite-reducing, anaerobic bacteria of *Thiobacillus* genus and moulds of *Penicillium*, *Acrodontium*, *Auerobasidium* and *Cladosporium* genera [Papciak and Zamorska 2007; Zyska 1999].
4) Concrete: As the foundations and walls of a building age as a result of contact with the atmosphere, soil, sewage, waste, chemicals etc., the concrete reaction changes to acidic, which favours microbial growth. When the environment is rich in sulphur compounds (e.g. hydrogen sulphide), such concrete is inhabited by sulphur oxidising bacteria, especially of the *Acidithiobacillus thiooxidans* genus [Cwalina and Dzierżewicz 2007]. On the other hand, a significant concentration of nitrogen compounds favours the growth of nitrifying bacteria such as *Pseudomonas*, *Proteus* or *Alcaligenes* [Książek 2014]. *Coniophora puteana* and *Serpula lacrymans* can also be responsible

for the biological destruction of concrete. Their impact may result in an increase in material moisture content by 18–25%, decrease in its pH from 12 to 5–7.7 and decrease in its strength by 5–20% [Zyska 1999].

5) Metals: The bacteria of the genera *Thiobacillus* (oxidising sulphur, thiosulphates, sulphides and other polysulphide compounds to form sulphates, causing the formation of acids), *Desulfovibrio* and *Desulfotomaculum* (oxidising iron sulphide to form sulphuric acid) as well as the iron-oxidising bacteria from *Gallionella* and *Sphaerotilus* genera play a major role in the initiation of corrosion of this type of material. Moreover, biofilm formed by bacteria of *Clostridium*, *Bacillus* or *Pseudomonas* genera [Beech and Gaylarde 1999] and by moulds of *Fusarium* and *Penicillium* genera [Kip and van Veen 2015] was found on corroded metal elements. Corrosion of metals is a dynamic phenomenon (its daily rate can be 100–885 $mg/dm^2$), which can lead to perforation of pipes and subsequently to water damage.
6) Paint coatings: The corrosion of this type of material mainly involves moulds of the genera *Aureobasidium*, *Alternaria*, *Cladosporium* and *Phoma*, which usually originate from atmospheric air. Their microbial growth depends on the availability of carbon and nitrogen sources in paints as well as the temperature and relative humidity of the air and substrate. Infected raw materials, easily decomposed by microorganisms, including vegetable oils, which are a nutrient for e.g. *Aspergillus niger*, *Aspergillus versicolor* and *Penicillium chrysogenum* as well as cellulose derivatives, which are a nutrient for e.g. *Aspergillus flavus* and *Stachybotrys chartarum*, contribute significantly to the spread of moulds on paints and varnishes [Papciak and Zamorska 2007]. Freezing of external structures, absence or blockage of ventilation ducts and insufficient heating of rooms encourage biodeterioration processes.
7) Products containing paper (wallpapers, plasterboards): The key factor here is the material moisture content – if it exceeds 10%, products of this nature can be colonised by moulds without any obstacles. Fungal growth on paper products is initially of a surficial nature, and later on, the hyphae penetrate the fibres and the phase of cellulose decomposition begins. The processes of cellulose decomposition are accompanied by the release of water, the production of mucus, musty odour and, above all, discolouration, which can help to distinguish the type of fungus inhabiting the surface. For example, pink or pinkish stains may indicate contamination by fungi of *Aspergillus*, *Chaetomium*, *Rhodotorula*, *Fusarium* or *Penicillium* genera; greyish brown stains indicate the presence of moulds of *Cladosporium* spp. and *Stachybotrys chartarum*; and black stains indicate the presence of *Aureobasidium pullulans* and *Phoma violacea* fungal species [Papciak and Zamorska 2007]. Physical and chemical properties of paper are irretrievably lost during bio-corrosion. No fungicides are used when it comes to wallpapers, and the only protection is mould prevention.
8) Glass: Despite its smooth structure, glass can also be susceptible to bio-corrosion. The main causative factors of this process are moulds, which cause glass erosions. First, such symptoms are caused by *Aspergillus*

*versicolor*. Then the species diversity increases and moulds of *Penicillium*, *Alternaria*, *Cladosporium*, *Botrytis* and *Fusarium* genera appear [Drewello and Weissmann 1997; Gutarowska 2014]. Biodeterioration of glass by fungi and their metabolites causes irreversible changes in its surface layers, which are manifested by discolouration and cracks.

9) Polyvinyl chloride (PVC): This is the basic raw material from which modern windows are produced. Due to the fact that such windows are characterised by high tightness, water vapour generated indoors can condense on their surfaces. Easy accessibility of water and carbon, which is present in PVC, may favour colonisation by both bacteria and fungi. *Pseudomonas aeruginosa* bacteria and yeast-like fungi *Aureobasidium pullalans* may appear first. Then (after approximately 60 weeks) other yeasts may appear (e.g. *Geotrichum candidum*) or moulds of *Aspergillus*, *Penicillium* and *Ulocladium* genera [Webb et al. 2000; Pečiulytė 2002].

## REFERENCES


Becker, R. 1984. Condensation and mould growth in dwellings: Parametric and field study. *Build Environ* 19(4):243–250.

Beech, I. B., and C. C. Gaylarde. 1999. Recent advances in the study of biocorrosion: An overview. *Rev Microbiol* 30(3):177–190.

Burge, H. A., ed. 1995. *Bioaerosols*. Boca Raton, FL: Lewis Publishers/CRC Press, Inc.

Ciferri, O. 1999. Microbial degradation of paintings. *Appl Environ Microbiol* 65(3):879–885.

Clausen, C. A., and V. Yang. 2007. Protecting wood from mould, decay, and termites with multi-component biocide systems. *Int Biodeterior Biodegrad* 59(1):20–24.

Cwalina, B., and Z. Dzierżewicz. 2007. Czynniki sprzyjające biologicznej korozji konstrukcji żelbetowych, Cz. 1. *Prz Budowlany* 7/8:52–59.

Dakal, T. C., and S. S. Cameotra. 2012. Microbially induced deterioration of architectural heritages: Routes and mechanisms involved. *Environ Sci Eur* 24(1):36.

Drewello, R., and R. Weissmann. 1997. Microbially influenced corrosion of glass. *App Microbiol Biotechnol* 47(4):337–346.

Ellringer, P. J., K. Boone, and S. Hendrickson. 2000. Building materials used in construction can affect indoor fungal levels greatly. *Am Ind Hyg Assoc J* 61(6):895–899.

Florian, M. L. E. 2000. Aseptic technique: A goal to strive for in collection recovery of moldy archival materials and artifacts. *J Am Inst Conserv* 1(1):107–115.

Gaylarde, C. C., and P. M. Gaylarde. 2005. A comparative study of the major microbial biomass of biofilms on exteriors of buildings in Europe and Latin America. *Int Biodeterior Biodegrad* 55(2):131–139.

Gooday, G. W. 1988. The potential of the microbial cell and its interaction with other cells. In *Micro-Organisms in Action: Concepts and Applications in Microbial Ecology*, eds. J. M. Lynch, and J. E. Hobbie, 7–32. Oxford: Blackwell Scientific Publications.

Gravesen, S., J. C. Frisvad, and R. A. Samson 1994. *Microfungi*. Copenhagen: Munksgaard.

Gutarowska, B. 2014. Moulds in biodeterioration of technical materials. *Folia Biol Oecol* 10(1):27–39.

Hyvärinen, A., T. Meklin, A. Vepsäläinen, and A. Nevalainen. 2002. Fungi and Actinobacteria in moisture-damaged building materials: Concentrations and diversity. *Int Biodeterior Biodegrad* 49(1):27–37.

Karbowska-Berent, J., and A. Strzelczyk 2000. *The Role of Streptomycetes in the Biodeterioration of Historic Parchment*. Torun: Wydawnictwo Uniwersytetu Mikołaja Kopernika.

Karunasena, E., N. Markham, T. Brasel, J. D. Cooley, and D. C. Straus. 2000. Evaluation of fungal growth on cellulose-containing and inorganic ceiling tile. *Mycopathologia* 150(2):91–95.

Kip, N., and J. A. van Veen. 2015. The dual role of microbes in corrosion. *ISME J* 9(3):542–551.

Kołwzan, B. 2011. Analiza zjawiska biofilmu: Warunki jego powstawania i funkcjonowania. *Ochrona Srodowiska* 33(4):3–14.

Książek, M. 2014. City sewer collectors biocorrosion. *Cent Eur J Eng* 4(4):398–407.

Kujanpää, L., S. Haatainen, R. Kujanpää, R. Vilkki, and M. Reiman 1999. *Microbes in Material Samples Taken from Base Boardings, Gypsum Boards and Mineral Wool Insulation. Edinburgh.* Scotland: IAIAS, 892–896.

Macher, J., eds. 1999. Bioaerosols: Assessment and control. *American Conference of Governmental Industrial Hygienists*, Cincinnati.

Nielsen, K. F., P. A. Nielsen, and G. Holm. 2000. Growth of moulds on building materials under different humidities. *Proc Healthy Buildings* 3:283–288.

Papciak, D., and J. Zamorska. 2007. Korozja mikrobiologiczna powodowana przez grzyby. *Zeszyty Naukowe Politechniki Rzeszowskiej. Budownictwo i Inżynieria Środowiska* 46(246):87–99.

Pečiulytė, D. 2002. Microbial colonization and biodeterioration of plasticized polyvinyl chloride plastics. *Ekologija* 4:7–15.

Peczyńska-Czoch, W., and M. Mordarski 1988. Actinomycete Enzymes. In *Actinomycetes in Biotechnology*, eds. M. Goodfellow, S. T. Williams, and M. Mordarski, 219–283. San Diego: Academic Press.

Pessi, A. M., J. Suonketo, M. Pentii, M. Kurkilahti, K. Peltola, and A. Rantio-Lehtimaki. 2002. Microbial growth inside insulated external walls as an indoor air biocontamination source. *Appl Environ Microbiol* 68(2):963–967.

Piontek, M., and H. Lechów 2013. Deterioracja Elewacji Zewnętrznych Wywołana Biofilmem. *Inżynieria Środowiska* 31, 77–85.

Reiman, M., L. Kujanpää, R. Vilkki, P. Sundholm, and R. Kujanpää. 2000. Microbes in building materials of different densities. *Proc Healthy Buildings* 3:313–316.

Saiz-Jimenez, C. 1994. Biodeterioration of stone in historic buildings and monuments. In *Biodeterioration Research 4. Mycotoxins, Wood Decay, Plant Stress, Biocorrosion and General Biodeterioration*, eds. G. C. Llewellyn, W. V. Dashek, and C. E. O'Rear, 587–604. New York: Plenum Press.

Schmidt, O. 2007. Indoor wood-decay basidiomycetes: Damage, causal fungi, physiology, identification and characterization, prevention and control. *Mycol Prog* 6(4):261–279.

Schmidt, O., and W. Kallow. 2005. Differentiation of indoor wood decay fungi with MALDI-TOF mass spectrometry. *Holzforschung* 59(3):374–377.

Sedlbauer, K. 2001. *Prediction of Mould Fungus Formation on the Surface of and inside Building Components.* Doctoral thesis. Stuttgart: Fraunhofer Institute for Building Physics.

Szczepanowska, H., and C. H. M. Lovett. 1992. A study of the removal and prevention of fungal stains on paper. *J Am Inst Conserv* 2(2):147–160.

Tuomi, T., K. Reijula, T. Johnsson et al. 2000a. Mycotoxins in crude building materials from water-damaged buildings. *Appl Environ Microbiol* 66(5):1899–1904.

Tuomi, T., T. Johnsson, and K. Reijula. 2000b. Mycotoxins and associated fungal species in building materials from water-damaged buildings. *Proc Healthy Buildings* 3:371–376.

Webb, J. S., M. Nixon, I. M. Eastwood, M. Greenhalgh, G. D. Robson, and P. S. Handley. 2000. Fungal colonization and biodeterioration of plasticized polyvinyl chloride. *Appl Environ Microbiol* 66(8):3194–3200.

Wołejko, E., and M. Matejczyk. 2011. Problem korozji biologicznej w budownictwie. *Budownictwo i Inżynieria Środowiska* 2:191–195.

Zyska, B. 1999. *Zagrożenia Biologiczne w budynku.* Warsaw: Arkady.

# 6 Methods of Identifying Microbiological Hazards in Indoor Environments

*Rafał L. Górny*

## CONTENTS



## 6.1 MICROBIAL AEROSOL SAMPLING METHODS

In each environment, the specific properties (physical and biological) of bioaerosol particles determine the method of their collection. In accordance with contemporary requirements for microbiological air pollution testing, it is recommended to use volumetric methods, which consist of sampling of a certain volume of the air. Techniques most frequently used for bioaerosol sampling include:

- impaction, in which the separation and collection of particles from the air stream onto a solid surface (e.g. microbiological medium) occurs by inertial forces;
- impingement, i.e. impaction into a liquid with particle diffusion within the liquid bubbles; this method is characterised by high physical and biological efficiency of particle collection;
- filtration, i.e. separation of particles during the air stream flow through a porous medium in the form of a filter; due to its simplicity, low cost and wide range of applications, filtration is a commonly used technique in this type of measurements;
- electrostatic precipitation, in which the separation takes place as a result of electrostatic interactions with charged particles suspended in the air; this method has a high efficiency and is considered promising and forward-looking due to the 'gentleness' of the particle collection process.

Bioaerosol sampling is designed to efficiently and effectively capture as many biological particles as possible from the air and then collect them in such a way as to enable their subsequent detection, i.e. without altering or damaging their structure while maintaining their ability to grow in or on a suitable microbiological medium. The adherence to these conditions depends on the physical and biological characteristics of the tested microorganism and on the physical collection efficiency of the measuring device used [Delort and Amato 2018; Després et al. 2012; Directive 2000/54/EC; EN 13098, 2002; EN 14031, 2002; EN 14583, 2005; Górny and Ławniczek-Wałczyk 2012; Hung et al. 2005; Kulkarni et al. 2011; Macher 1999; Yang and Heinsohn 2007].

All the above-mentioned bioaerosol sampling methods have their advantages and disadvantages. According to the recommendations of the American Conference of Governmental Industrial Hygienists, considerable flexibility in choosing the method of microbiological analysis of the air is allowed on condition that it ensures the repeatability and reliability of the results. A similar position is represented by the experts from the European Union, who specify that in the assessment of the hygienic quality of indoor spaces, both the concentration and qualitative composition of microbiota should be determined [Delort and Amato 2018; Directive 2000/54/EC; EN 13098, 2002; Kulkarni et al. 2011; Macher 1999].

Compared to other aerosols, bioaerosols require specific sampling procedures. Traditional sampling methods (i.e. filtration and impaction, including impingement into a liquid) and their analyses (culture methods) are aimed at the evaluation of viable microbial particles such as conidia, spores and vegetative cells. When using these methods, however, the importance of particles unable to grow and form colonies (i.e. viable but non-culturable microorganisms), non-viable microorganisms and their fragments are usually not taken into account. The consequences of such an action, i.e. measuring only a small proportion of the particles present in a given environment, is a conscious underestimation of the real exposure caused by biological agents. Both the quantitative and qualitative results of the analysis carried out using the above-mentioned methods (whether they are carried out as stationary or personal measurements) can be influenced by a number of interfering factors. These include, for example, sampling time (which is usually relatively short) or numerous environmental, spatial and temporal variations. Even the use of microbial source characteristics, usually carried out by taking samples from a microbiologically contaminated surface or, less frequently, by assessing the microbial source strength (based on techniques that use forced air movement to maximise the aerosolisation of microbial particles), may result in inaccurate estimation of their emissions into the air [Crook 1995a, 1995b; Delort and Amato 2018; Després et al. 2012; Hung et al. 2005; Kulkarni et al. 2011; Macher 1999; Yang and Heinsohn 2007].

Sampling instruments using inertial forces are widely used in microbiological aerosol measurements. The most commonly used are single- and multi-stage (cascade) impactors and slit samplers (if a given stage of impactor consists of slits instead of circular holes). Cascade impactors (six-, seven-, eight- or ten-stage) especially are used as reference devices for collecting viable microorganisms that can grow on culture media used according to their taxonomic origin. These instruments ensure efficient separation of the particles collected along with the air stream aspirated by the sampler, due to a precisely defined 'cut-off diameter' of particular impactor's stage. Impactors of this type are available for both stationary (e.g. Andersen impactor) as well as individual (e.g. Sioutas impactor) measurements (Figure 6.1 and Figure 6.2). On the other hand, impingers (such as AGI-30 or BioSampler – Figure 6.3) combine two mechanisms of airborne particle collection, i.e. impaction into the liquid and diffusion of particles within the bubbles of the capturing fluid. In comparison to impactors, impingers allow for a significant extension of sampling time, while maintaining both high microbial particle collection efficiency and their biological properties. In turn, electrostatic (Figure 6.4) or thermal precipitators, which enable 'gentle' microbial aerosol particle collection, while maintaining its viability

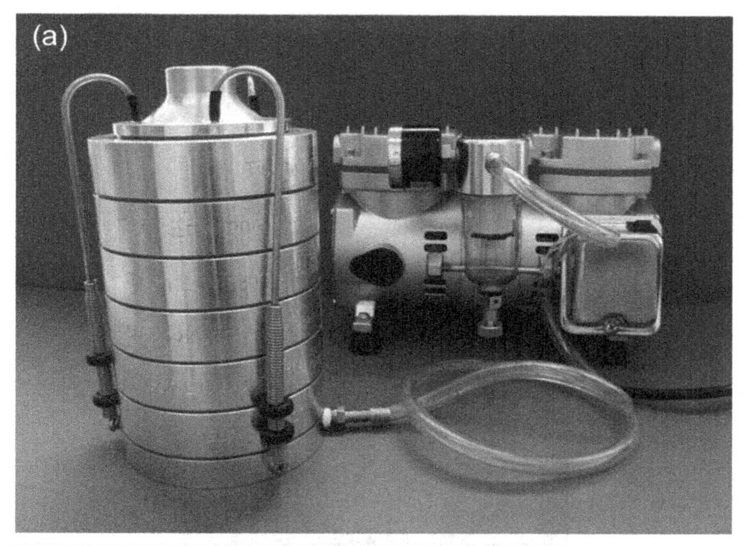


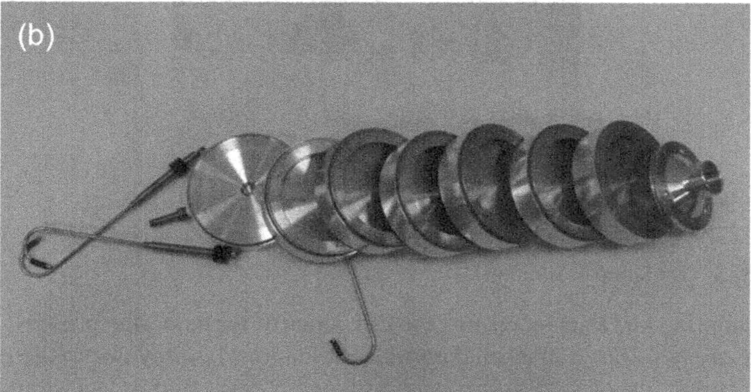


**FIGURE 6.1** Six-stage Andersen impactor: (a) measurement set with a pump and (b) impactor components (left: base, six impactor stages, inlet).

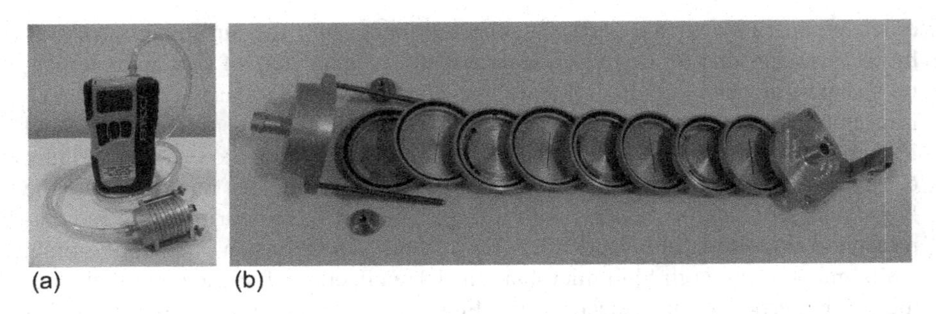
(a) (b)

**FIGURE 6.2** Sioutas: (a) general view and (b) impactor components (left: inlet with cover, base, impactor stages).

**FIGURE 6.3** BioSampler impinger.

during sampling and thus increasing the efficiency of the sampling process, are used relatively rarely so far in microbial particle evaluations (mainly due to the low commercial availability of such instruments) [Crook 1995a, 1995b; Delort and Amato 2018; Després et al. 2012; Hung et al. 2005; Kethley et al. 1952; Kulkarni et al. 2011; Macher 1999; Mainelis et al. 1999, 2001, 2002; Orr et al. 1956; Tan et al. 2011; Yang and Heinsohn 2007].

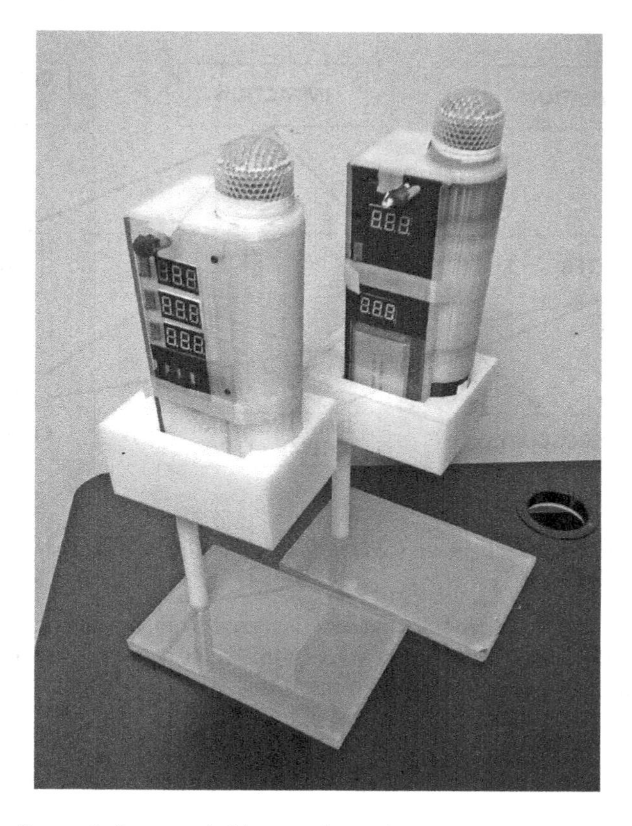

**FIGURE 6.4** Personal electrostatic bioaerosol samplers.

## 6.2 SAMPLING METHODS FOR MICROBIOLOGICALLY CONTAMINATED SURFACES

Methods used for sampling from microbiologically contaminated surfaces include:

- imprinting using an adhesive tape (tape is pressed to a contaminated surface and subsequently directly transferred to the microscope slide and analysed); suitable for flat and smooth surfaces;
- swabbing (sterile swab is moistened with an adequate liquid – sterile water, saline, peptone water etc. – and used to collect microorganisms deposited on a contaminated surface, which are then suspended in a liquid of a larger volume, usually identical to the one used to take the sample, and analysed with serial dilution method); suitable for corrugated and porous surfaces;
- contact plates (special RODAC plates are usually used, filled with a suitable culture medium, forming a convex meniscus, with a contact surface of not less than 20 cm$^2$, which are pressed against the contaminated surface for several seconds; sometimes an agar tape is used for the same purpose); suitable for flat and smooth surfaces;

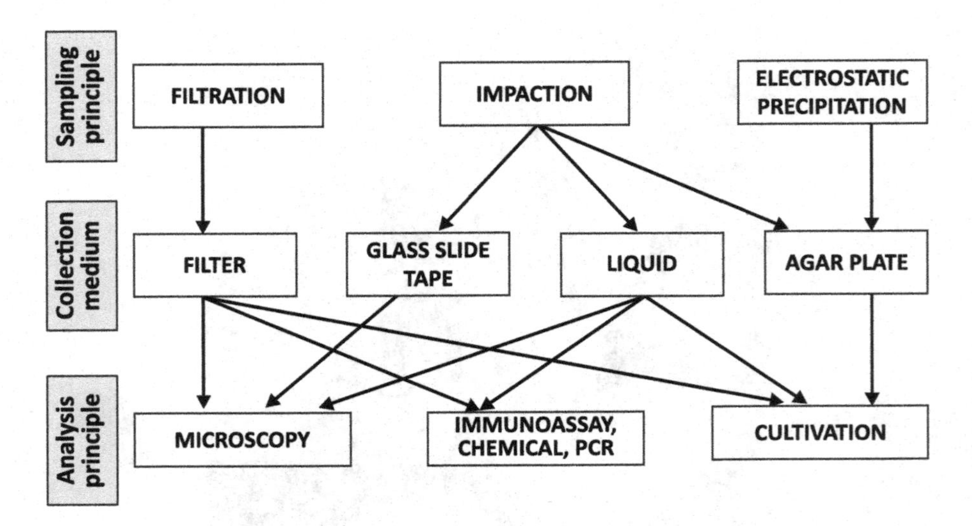


**FIGURE 6.5** Interactions between techniques for collecting and analysing environmental microbiological samples.

- vacuuming (the microbial material is taken with a vacuum cleaner on disposable – usually cotton – filters, on which it is first gravimetrically assessed and then analysed with the serial dilutions method); suitable for any type of surface.

The choice of appropriate method depends on the type of analysed surface and the further planned phases of analytical sample preparation. Figures 6.5 and 6.6 represent correlation diagrams between the methods of collection and analysis of microbiological samples and the algorithm of analytical procedures used in the quantitative and qualitative evaluations of microbiological pollutants [Bakal et al. 2017; Hung et al. 2005; Macher 1999].

## 6.3 QUANTITATIVE AND QUALITATIVE ANALYSIS OF MICROBIOLOGICAL SAMPLES

In order to determine the concentration and taxonomical composition of microorganisms present in environmental samples, the following methods are used:

- microscopic examination, consisting of assessment of the number of microbial cells and then calculation of their concentration in an air volume unit or on a tested area – the advantage here is that all microorganisms are recorded (i.e. viable and non-viable), and the disadvantage is that it is not possible to precisely identify the taxonomical affiliation of the species;
- cultivation method, which enables determining the number of viable (understood here as culturable) microorganisms; their concentration is expressed as colony forming units, CFU, an air volume unit or on a tested area; this

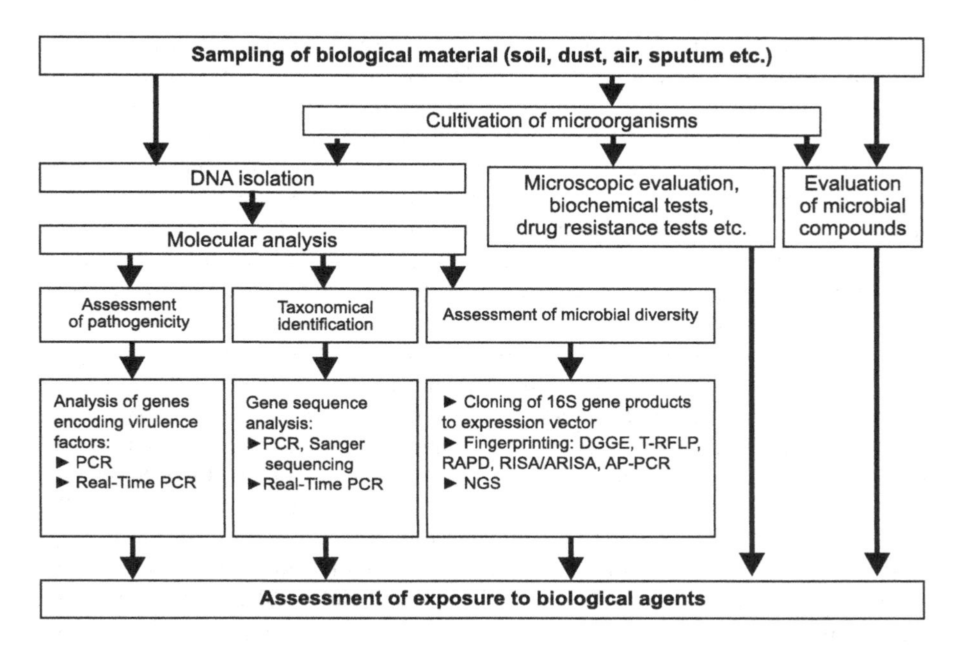


**FIGURE 6.6** Analytical algorithm for the evaluation of environmental samples with regard to the quantitative and qualitative assessment of microbiological contamination in them.

method enables identifying isolated microorganisms up to genus and/or species level;

- metabolic and molecular profiling, in which the concentration of microorganisms in the air or on a contaminated surface is determined on the basis of the presence of their metabolic products, non-specific and specific DNA or gene probe; quite frequently immunologic tests based on the use of mono- and polyclonal antibodies or markers, including immunological markers, that reveal the presence of specific groups or strains of microorganisms are also used; molecular biology techniques are also often used to identify individual strains of microbial species.

Traditional techniques widely used to characterise taxonomic diversity and quantitative presence of microorganisms in the air and surface samples are usually based on culture methods and accompanying microscopic analyses. Although these methods may be considered somewhat 'obsolete', they are still successfully applied in research today. However, it should be emphasized that, despite their utility, these methods have considerable limitations. This is because they are dedicated to detection of viable (usually understood as culturable) microorganisms, and thus omit remaining microbial particles, both viable but non-culturable under laboratory conditions, and non-viable, as well as their fragments. It is worth mentioning here that most microbial particles present in the air, even if they remain viable, are unable to reproduce and form new colonies, even on appropriately prepared media. This is why conclusions drawn from air pollution tests which use cultivation methods significantly reduce the value of real exposure. Results obtained thus far indicate that the number of viable microbial particles able to grow as separate colonies on a suitable microbiological medium is not higher than 25% of all airborne bacteria and 17% of all airborne fungi; in fact, when all airborne microorganisms are considered, these values are typically a lot lower, between 0.03% and 1%. Factors which influence this state of affairs include those of technical (methods and time of sampling, type of medium, incubation conditions), biological (strain of microorganism, type of particles collected, i.e. vegetative cells, spores, conidia) and environmental (including temperature, relative humidity, time of day, season, climate zone, geographical region, types of natural reservoirs such as water, soil, forests, deserts, presence of plant and animal populations etc.) origin [Amann et al. 1995; Amato et al. 2007; Bridge and Spooner 2001; Chi and Li 2007; Colwell 2000; Cox and Wathes 1995; Dutkiewicz and Jabłoński 1989; Gołofit-Szymczak and Górny 2010; Griffin et al. 2006; Heidelberg et al. 1997; Hung et al. 2005; Lighthart 2000; Rappé and Giovannoni 2003; Roszak and Colwell 1987; Shahamat et al. 1997; Staley and Konopka 1985; Stewart et al. 1995; Tong and Lighthart 1999; Wainwright et al. 2004; Wang et al. 2001, 2007, 2008; Yang and Heinsohn 2007].

Ever since the first optical microscope was built in the 16th century, microscopic techniques have played an important role in microbiological research, including research related to bioaerosols. They allow one to observe the structure and determine the dimensions of both large (micrometric) and smaller (submicron and nanometric) microorganisms. Microscopes not only make possible the quantitative identification of viable and non-viable microorganisms, but also enable assessment of the number of such factors, starting from microbial vegetative cells, spores or conidia, through small infectious factors, i.e. viruses, and ending with molecules which constitute elements of their cellular structure. Light or bright-field microscopes are typically used for simple observations of shapes, sizes or the number of microorganisms.

Particles in samples can be observed in transmitted light (for transparent specimens) or reflected light (for opaque specimens), in light field or dark field or in polarised light to enhance the contract and precisely visualise the details of observed elements. Specimens can be stained using e.g. methylene blue, crystal violet, safranin, fuchsine or other differentiating stains in order to classify microbiological particles into appropriate groups. On the other hand, when the observed particles are almost invisible and observation of the specimen in another medium is impossible or unacceptable, a phase-contrast microscope is used. Fluorescence microscopy is a variant of light microscopy; it uses an ultraviolet or near-ultraviolet light source which causes particles with fluorescent properties to emit light.

Direct particle counting methods used e.g. to determine the total number of microorganisms (i.e. both viable and non-viable) are one of the most commonly used analytical techniques in aerobiological research today. Using various types of fluorochromes (e.g. acridine orange; 4′,6-diamidino-2-phenylindole; 5-cyano-2,3-ditolyl tetrazolium chloride; fluorescein isothiocyanate), either selectively binding to structural elements of examined cells or paired with appropriate antibodies or molecular markers, makes it possible to locate specific cellular structures or observe selected physiological processes.

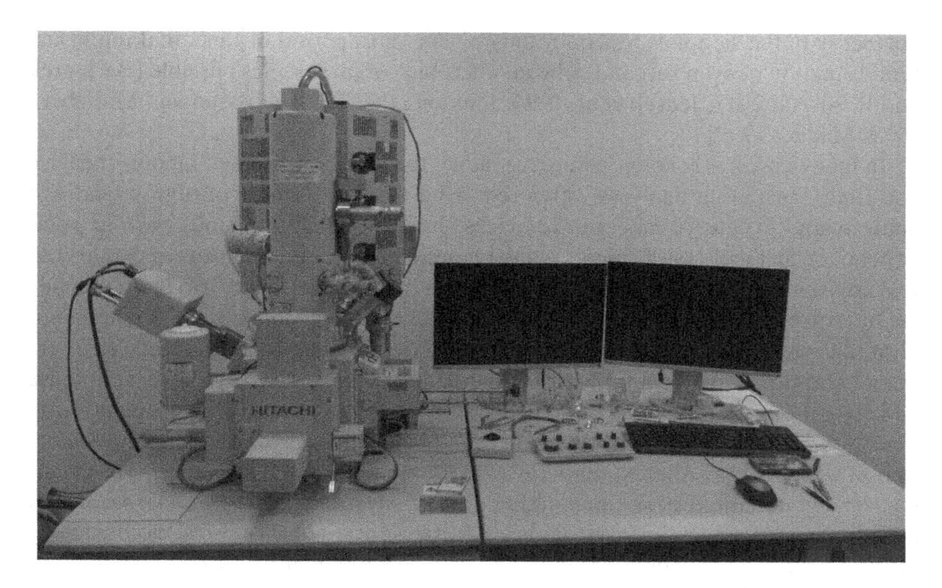


**FIGURE 6.7** Scanning electron microscope (SEM).

Electron microscopy, in turn, uses a beam of electrons instead of light. In a scanning electron microscope (SEM), the surface of the sample is scanned using a collimated electron beam (Figure 6.7). Electron signals obtained from its interaction with the surface of the examined sample are received by appropriate detectors and converted into images. A high-vacuum operation mode is the primary mode for all modes of SEM operation. It makes it possible to magnify the sample up to 10 million times its original size. In aerobiological research, SEM is typically used to determine the size and shape of individual microbiological particles and assess their ability to form aggregates and bind to other aerosol particles or fibres [Bartoszek and Rosowski 2017; Després et al. 2012; Francisco et al. 1973; Harrison et al. 2005; Hernandez et al. 1999; Hobbie et al. 1977; Jensen et al. 1994; Karlsson and Malmberg 1989; Kepner and Pratt 1994; Macher 1999; Pöhlker et al. 2011].

In the case of a transmission electron microscope (TEM), a beam of high-energy electrons produced by the electron gun is directed at a thin (e.g. several dozen or several hundred nanometres thick) sample, at which point it can be reflected, absorbed or transmitted. In transmission mode, reflected electrons are used to create an image of the sample structure, while in scanning mode, they image its surface. TEM has a resolution limit approaching 0.1 nm, which allows observation of even the atomic structure of the examined sample. Going further, an atomic force microscope (AFM) utilises a scanning probe to examine the surface by recording the forces which act on it as a function of position. AFM analyses deflection of a cantilever with a tip under the influence of interaction forces (primarily van der Waals forces) between the atoms of the tip and the atoms which make up the examined surface. A detector (a photodiode) converts the deflection of the cantilever into a current signal, which is then used to generate the sample image. In pulsed mode, the probe tip remains in contact with the sample surface for only a very short period of time, making visual representation of even soft and delicate microbiological samples possible [Bartoszek and Rosowski 2017; Jensen et al. 1994; Jonsson et al. 2014; Karlsson and Malmberg 1989; Macher 1999].

In many cases, it is not possible to conduct the above-mentioned labour-intensive and time-consuming analyses, often requiring the involvement of highly qualified analysts and expensive laboratory devices. In such cases, numerous cellular components or metabolites of microorganisms may be temporarily measured to reflect real environmental exposure to harmful agents of microbial origin. So far in practice we have used several chemical markers in such cases e.g. endotoxins as markers of Gram-negative bacteria contamination, muramic acid from peptidoglycans as markers of Gram-positive bacteria contamination, ergosterol, N-acetyl-hexosaminidase and (1-3)-β-D-glucans as markers of fungal biomass. A number of instrumental and bioanalytical techniques have also been tested in practice, allowing for precise detection of the above-mentioned markers of microbiological contamination. For example, for the quantitative and qualitative evaluation of endotoxins in environmental samples, a wide range of *in vitro* analyses is used. They include several modifications of the *Limulus* amoebocyte lysate (LAL) test, recombinant factor C (*rFC*) assay or liquid and gas chromatography (used separately or in combination with mass spectrometry, GCMS). In turn, tests of muramic and diaminopimelic acids as markers of peptidoglycans, which are components of the bacterial cell wall structure, can be performed using GCMS.

The detection of β-glucans can also be performed with four different analytical techniques: modified version of the LAL test, the enzyme immunoassay (EIA) and the enzyme-linked immunosorbent assay (ELISA) as well as the test using monoclonal antibodies (mAbs) and immunoenzymatic inhibition of mAb-EIA. Markers characterising fungal biomass include ergosterol estimated quantitatively by GCMS, extracellular polysaccharides analysed by specific immunoenzymatic reactions and N-acetyl-hexosaminidase, the activity of which is assessed by the method based on fluorescently marked substrate decomposed by the enzyme present in fungi. Furthermore, the products of fungal secondary metabolism, i.e. mycotoxins, can be analysed using different variants of chromatography: thin-layer chromatography, gas chromatography (with or without mass spectrometry) or high-performance liquid chromatography (HPLC). The latter two techniques can also be used for quantitative analyses of microbial volatile organic compounds (MVOCs), which can serve as markers of fungal growth in the environment. The analysis of advantages and disadvantages of the above-mentioned methods clearly shows that the assessment of chemical markers of microbiological contamination, despite providing precise quantitative information, does not give information on the biodiversity of microbiological particles. Fortunately, this 'knowledge gap' can be filled with the help of a whole spectrum of molecular biology techniques that enable the precise identification of bioaerosol-forming microbial strains (see Section 6.4.3) [Demirev and Fenselau 2008; Després et al., 2012; Douwes et al. 1999; Griffith and DeCosemo 1994; Hung et al. 2005; IOM 2004; Macher 1999; Miller and Young 1997; Pöschl 2005; Reeslev et al. 2003; Reponen et al. 1995; Rylander et al. 2010; WHO 2009].

Since none of the above-mentioned methods offers real-time bioaerosol detection, various optical techniques are available to overcome this limitation. Among them are those that use scattering and concentration of light and fluorescence of microbiological particles under its influence. Some of these methods offer both measurements and separation of microbiological particles, starting with those of nanometric sizes. Among the devices using the principles described above are optical particle counters (using the phenomenon of light beam scattering e.g. Grimm 11A; enabling the measurement of optical diameters of particles in the range of 0.25–32 µm), condensation counters, counting particles on which alcohol vapours have condensed (e.g. P-TRAK giving the possibility of counting particles with diameters of 0.01–~2 µm), aerodynamic spectrometers (combining the possibility of analysing the aerodynamic diameters of particles and the intensity with which they scatter light e.g. DSP Aerosizer, enabling the measurement of particles in the range of 0.2–200 µm) and scanning mobility particle sizers (consisting of a condensation particle counter and differential mobility analyser, measuring particle concentration and giving their size distribution e.g. SMPS, possibility of analysing particles of 1–1000 nm diameter) (Figure 6.8). In turn, the electrical low pressure impactor (ELPI) combines the quantitative control of microbiological particles with their real-time size distribution characteristics (in the range of 0.006–10 µm). In ELPI, as in many other volumetric instruments, the sampling process is inextricably linked to certain measurement limitations. In this device, particles hitting against hard surfaces, then bouncing off them, and consequently their re-aerosolisation, friction during the passage through the instrument and drying, can significantly reduce the accuracy of measurement and control of bioaerosol concentration. These phenomena should always be taken into account when sampling of biological particles with volumetric instruments is performed [Hung et al. 2005; Jonsson et al. 2014; Kulkarni et al. 2011; Macher 1999].

The operation of devices using fluorescence as a detection method is based on the phenomenon of all biological particles having fluorophores, derived from amino acid residues (intracellular components of almost all proteins). Tryptophan, tyrosine and phenylalanine are amino acids capable of fluorescence when stimulated by UV light, and this phenomenon can be used to detect microbial aerosols. One of the first commercial applications of induced fluorescence was creating the fluorescence aerodynamic particle sizer (FLAPS) and ultraviolet aerodynamic particle sizer (UV-APS) (Figure 6.9). Both devices are designed for real-time bioaerosol analysis. FLAPS technology in its latest version uses a CW laser diode – both for fluorescence excitation and optical measurements. The system analyses individual particles in the size range of 0.8 µm to 10 µm and differentiates microbial aerosol particles including bacteria, fungi, viruses and toxins. In turn, UV-APS (which is technically a newer version of FLAPS) measures the fluorescence of aerosol particles in real time (after their excitation with a pulsed UV laser). This device simultaneously provides information on grain size distribution (in the range of 0.8–15 µm) and intensity of light scattering on aerosol particles. It has so far been used mostly to investigate sources of bioaerosol emissions, to study particle properties in laboratory tests, clean rooms, hospitals and other interiors (including industrial and residential environments) and to control the atmosphere in urban and rural conditions. Another meter in this group

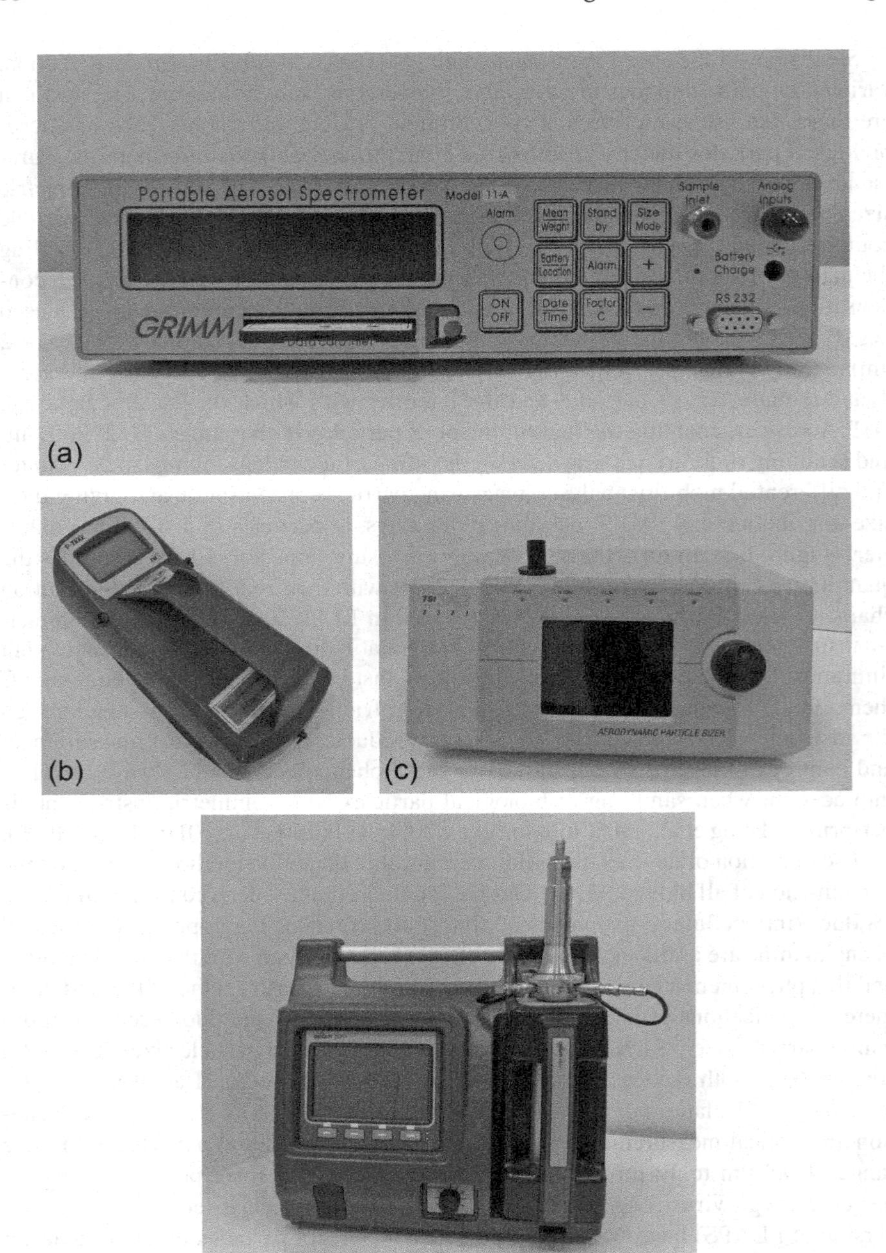


**FIGURE 6.8** Real-time particle counters: (a) GRIMM optical particle counter, (b) P-TRAK condensation particle counter, (c) Aerosizer and (d) ELPI electrical low pressure impactor.

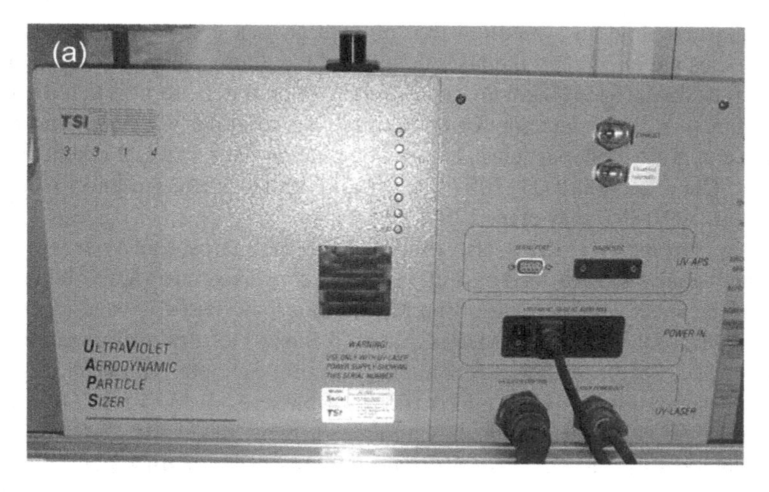

**FIGURE 6.9** Fluorescent particle classifiers: (a) ultraviolet aerodynamic particle sizer (UV-APS) (TSI.com) and (b) wideband integrated bioaerosol sensor (WIBS) (dropletmeasurement.com).

of measuring devices is the wideband integrated bioaerosol sensor (WIBS), which allows simultaneous optical detection and sizing of aerosol particles using a continuous wave CW laser and fluorescent detection, resulting from a pulsating UV light produced by two xenon lamps. During measurements, the device collects five different types of information, including optical particle size, particle asymmetry and fluorescence using three channels. WIBS in one second can process information about 125 particles and has been successfully tested in urban and tropical environments [Brosseau et al. 2000; Delort and Amato 2018; Després et al. 2012; Eng et al. 1989; Gabey et al. 2011; Harrison and Chance 1970; Hairston et al. 1997; Iwami et al. 2001; Jonsson and Kullander 2014; Jonsson et al. 2014; Kaye et al. 2005; Kell et al. 1991; Li et al. 1991].

Flow cytometry, similarly to epifluorescence microscopy, is nowadays used successfully to control in real time the presence of microbial particles in the air. Such particles may emit fluorescence naturally or under the influence of specific staining.

During laminar flow, in a medium free of such particles (e.g. in deionised water), such particles are excited by the laser beam, and the fluorescence emitted as a result and the light scattered on them are measured by photon detectors. When it comes to microorganisms, these detectors can determine e.g. total particle count and number of viable and non-viable particles and provide data on their size and taxonomic classification [Chen and Li 2005, 2007; Delort and Amato 2018; Ho and Fisher 1993; Lange et al. 1997; Prigione et al. 2004].

Mass spectrometry (MS) is also used in microbial aerosol analysis. In order to enable such analysis, microbial particles must be converted into vapour before being placed in the spectrometer. One of the MS methods successfully used in microbial studies is MALDI-TOF MS – Matrix Assisted Laser Desorption/Ionization Time-of-Flight Mass Spectrometry. MS can analyse samples containing both large numbers of microorganisms and single particles of bioaerosol. MALDI-TOF identifies microorganisms by determining their unique protein 'fingerprints'. Characteristic spectral patterns of microorganisms are used to precisely identify them by comparing and matching them with the spectra of reference strains. Both fast detection and accurate identification have determined the usefulness of MS techniques in the study of bacterial and fungal contamination present in various environments in the form of microbial aerosol [Jonsson and Kullander 2014; Kim et al. 2005; Kleefsman et al. 2008; Madsen et al. 2015; Van Wuijckhuijse et al. 2005].

## 6.4 MOLECULAR TECHNIQUES IN MICROBIOLOGICAL DIAGNOSTICS

Over the last two decades there has been a dynamic development of molecular methods in microbiological diagnostics. The most popular in this area are techniques based on the polymerase chain reaction (PCR), which allows obtaining a practically unlimited number of copies of any fragment of the genome in a relatively short time. These techniques are characterized by high sensitivity and specificity and are not dependent on the conditions of culturing on microbiological media. They are based on analysis of the genetic material of the microorganism, which is unique and not influenced by environmental changes, as are the phenotypic characteristics. Molecular techniques enable identification of microorganisms that are difficult to culture and isolate in laboratory conditions and allow for their precise and unambiguous species identification, and strain differentiation within a certain species [Alvarez et al. 1994; Blais-Lecours et al. 2012; Mukoda et al. 1994; Peccia and Hernandez 2006; Prussin et al. 2014].

### 6.4.1 Microorganism Species Identification

In the PCR, the adequate design of the primers allows for amplification of the genome regions responsible for specific metabolic characteristics of the microorganism e.g. antibiotic resistance, ability to produce toxic metabolites or other virulence factors such as adhesives, haemolysins or cytotoxic compounds. Therefore, this method allows not only assessment of the presence of a certain species of the microorganism in the test material, but also determination of its metabolic profile and pathogenic potential. In the identification of microorganisms isolated from the air, due to their usually low concentrations (approx. $10^3$–$10^4$ CFU/m$^3$), the so-called nested-PCR, in which two reactions are carried out sequentially, is used. In the first one the primers are designed to amplify relatively long (approx. 1000–1500 base pairs, bs) fragments. After purification, it becomes the matrix for the second reaction which product is nested within the first sequence. This approach allows an increase in sensitivity of the method as much as ten thousand times [Leema et al. 2011; Ramachandran et al. 2007; Sanger and Coulson 1975; Yamamoto et al. 1995].

Sequencing is a technique that allows determination of the sequence of nucleotides in the analysed DNA fragment. Its development in the 1970s by Frederick Sanger initiated extensive research aimed at creating genome maps of living organisms. The most famous of the research carried out was completed in 2003: the Human Genome Project. Currently, there is a very extensive, publicly available base that contains genome sequences of almost 260 thousand organism species and various bioinformatics tools for their analysis '*in silico*'. Although in recent years the interest in new sequencing techniques, such as pyrosequencing or next-generation sequencing, is growing, Sanger sequencing is still considered a reference method. It is also still the most commonly used technique to determine nucleotide sequences in the tested fragment of nucleic acid [Benson et al. 2013; Stärk et al. 1998].

Identification of the microorganism by PCR requires an amplification of that region of the genome which sequence is unique to the given species. The most commonly used 'molecular targets' of this type are the RNA coding genes of a small subunit of the ribosome e.g. 16S rRNA for bacteria and archaeons; 18S, 28S rDNA or fragments located between them, called internal transcribed spacers (ITS) for fungi. These genes, due to their structure, are called molecular chronometers. They are defined by the presence of both preserved (slowly changing in the course of evolution) fragments and those characterised by inter-species variability [Blais-Lecours et al. 2012; Chakravorty et al. 2007; Nehme et al. 2008; Oppliger et al. 2008; Perrott et al. 2009; Verreault et al. 2011; Woese 1987].

Analysis of 16S rRNA gene coding sequences of small ribosomal subunits (SSU) has become a globally accepted strategy to test phylogenetic differentiation of microorganisms that allows determination of species affiliation. Due to its high sensitivity and ability to identify rarely isolated, poorly characterised or phenotypically variable species, it is an excellent alternative to biochemical tests (e.g. API tests). The possibility of applying SSU rRNA gene sequence analysis has been significantly useful for differentiating clinically important pathogens, and also for identifying species of microorganisms whose detection by culture techniques is long lasting e.g. *Helicobacter* spp. or *Mycobacterium* spp., which identification by growth tests takes 6–8 weeks [Bosshard et al. 1996, 2006; Clarridge 2004; Henry et al. 2000; Kox et al. 1995; Nilsson et al. 2000].

### 6.4.2 Quantitative Analysis

The identification technique based on classical PCR and sequencing, although enabling species identification, does not allow determination of the number of microorganisms in the test material. A quantitative PCR technique (qPCR, Real-Time

PCR), because of the use of fluorescently marked oligonucleotides (called molecular probes) and adequate detection systems, allows monitoring of the increase in the number of product copies during the reaction. The advantage of this modification is both the possibility of detecting a specific DNA sequence in the test material, and also of quantifying it (although it is possible only if, in parallel with the carried-out reaction, we analyse samples with a known number of copies of the tested sequence, which allows plotting of the curve). Their adequate selection is crucial for quantitative analysis and may pose a challenge both in technical and scientific terms for laboratories studying the microbiological quality of the environment [Hospodsky et al. 2010; Nadkarni et al. 2002]. In recent years, the qPCR technique has been increasingly used in aerobiology to estimate the total count of bacteria, fungi or archaeons. Its use in bioaerosol studies in various environments showed that the number of microorganisms determined on the basis of these techniques was 100–1000 times higher than that determined by culture techniques [Blais-Lecours et al. 2015].

### 6.4.3 Microbial Population Diversity Analysis

Precise evaluation of exposure to harmful microbiological agents should include a quantitative and qualitative analysis of the population of microorganisms present in a certain environment. Analysis of 16S rRNA coding gene enables species identification, but for that it requires a clean culture of the tested organism beforehand. If this condition is not fulfilled, the reaction product will be a mixture of fragments – of the same size, but different sequences – of different microorganisms from the test material. Separation of these products by classical electrophoresis is not possible. To address this shortcoming, several research strategies have been elaborated for enabling an evaluation of the composition of mixed microbial populations based on the analysis of PCR products of 16S rRNA coding genes. These include cloning and analysis of gene libraries, techniques called 'fingerprinting' and next-generation sequencing.

Gene library development involves the cloning of a PCR product of the coding 16S rRNA which is amplified from the total DNA existing in the biological material to an adequate expression vector (e.g. pGEM-T). Bacteria are transformed by the library prepared in this way. After their incubation, individual colonies are isolated and each of them carries a vector with a fragment from one microorganism present in the starting material. The sequencing of vector fragments and their bioinformatics analysis allows determination of the species diversity of microbial populations in the examined environment [Blais-Lecours et al. 2011, 2014; Ma et al. 2015; Nehmé et al. 2009; Paez-Rubio et al. 2005].

The standard technique of PCR product analysis is electrophoresis in agarose or polyacrylamide gel. It enables the separation of DNA fragments of different sizes, using the differences in the speed of their migration in a gel placed in a constant electric field. It is commonly used for preliminary, qualitative analysis of PCR products as it allows determination of the length of the resulting product and the specificity of the primers used. The problem in the analysis of 16S rRNA amplification products after the PCR reaction is, despite differences in nucleotide sequences of different organisms, the identical length of the resulting products. In this case, classical electrophoretic analysis does not allow for their separation. In 1993, a technique was developed in which the separation of nucleic acid fragments is carried out in a polyacrylamide gel in the gradient of denaturing agent concentration, which is called denaturing gradient gel electrophoresis (DGGE). In this technique, the differences in the rate of denaturation (melting) of DNA fragments with different sequences, resulting from hydrogen bond strength difference between different nucleotide pairs, were used. DGGE analysis of the 16S rDNA PCR product allows determination of the phylogenetic diversity of the microbial population in a given environment by obtaining a characteristic profile (complex pattern) of differentiation of the 16S rRNA (for bacteria), 18S rRNA or ITS fragment (for fungi) coding gene sequence. This profile is called a 'molecular fingerprint', because it reflects the species composition of the microbial population. The complexes obtained through electrophoresis can be cut out of the gel, cleaned and sequenced for the purpose of identifying the microorganism. The sensitivity of such a technique is high, which even allows the detection of a microorganism that makes up only to 1% of the population. Due to the fact that DGGE is a semi-quantitative technique (differences in the intensity of complexes in the gel depend on DNA concentration in the starting material), it allows determination of the dominant species of microorganisms in a certain environment [Duarte et al. 2010; Gandolfi et al. 2013; Madsen et al. 2015; Maron et al. 2005; Muyzer et al. 1993; Tanaka et al. 2015].

In addition to DGGE analysis, other fingerprinting techniques have been developed to allow a rapid evaluation of microbiological biodiversity of the examined environment. They are mainly used to monitor changes in it or to compare microbial populations of different environments. And so in the ribosomal intergenic spacer analysis (RISA) technique, and its automated modification, ARISA (automated RISA), in a polyacrylamide gel the dimensions (lengths) of PCR product of a genome fragment, located between the genes coding the small (16S) and large (23S) ribosome subunit (intergenic spacer region, IGS), are analysed. The length of this fragment is characterised by the very high inter-species variability, varying between 150–1500 bs. The use of fluorescently marked primers and the automation of electrophoresis (ARISA) significantly increase the sensitivity of the method [Fisher and Triplett 1999; Pillai et al. 1996].

The terminal restriction fragment length polymorphism (T-RFLP) is an electrophoretic analysis of fluorescently labelled PCR 16S rRNA products, preceded by their digestion with an adequately selected restriction enzyme, crossing the double strand of DNA between nucleotides within a characteristic, several-nucleotide sequence (usually 6–8 bs). The only fragments separated here are those in which in the course of evolution were revealed such mutations which have resulted in the appearance of new particular enzyme-restricted zones. The restrictive fragments profile obtained reflects the relative diversity of microorganisms in the test material. The T-RFLP method is used to analyse populations originating from different environments, including those present in the air [Diel et al. 2005; Lee et al. 2010].

The arbitrarily primed polymerase chain reaction (AP-PCR) is a method for comparing the similarities between microbial strains within one species. It uses a single primer PCR technique and the product is separated in agarose gel. The AP-PCR method is very sensitive and allows tracking the spread of microbial strains in the environment [Orsini et al. 2002].

In the random amplified polymorphic DNA (RAPD) method, the PCR reaction is carried out with a short primer (8–12 nucleotides), which connects to the matrix at random. Its advantage is that the knowledge of the gene sequence of the analysed strains is not required. It has been used, inter alia, in genotyping clinically important species of mould and bacteria that are difficult to distinguish by culture methods. It has been proven that potentially pathogenic bacteria can be transmitted from the working environment to the hands and to masks which were used by workers as respiratory protection [Kermani et al. 2016; Ławniczek-Wałczyk et al. 2017].

The advantage of the methods generally referred to as fingerprinting is the possibility of quickly obtaining information on the differentiation of microorganisms in the examined environment without the need to create a microbial culture. Their limitations include poor repeatability (depending on many factors such as quality of isolation, sequence of primers used, PCR reaction conditions or electrophoresis conditions), lack of ability to identify microorganisms that are a small percentage of the population or poor ability to distinguish closely related microorganisms.

The development of molecular biology and advanced bioinformatics analysis techniques has enabled the development of new sequencing methods that allow the identification of the nucleotide sequence of increasingly longer DNA fragments in an increasingly shorter time, with a significant workload and costs reduction. Nowadays, the next-generation sequencing (NGS) technology allows obtaining a sequence of DNA fragment of a length from 3 to as much as 100 billion base pairs in about 24 h. This is possible due to, among other things, the application and improvement of a strategy called 'shotgun sequencing' (sequencing many short overlapping nucleic acid fragments of 50–500 bs length and then assembling them to obtain the entire output sequence) and the availability of an advanced detection system and software capable of processing huge amounts of data in a short time. Although this method is not yet widely used in routine analysis of exposure to harmful microbiological agents, it is worth emphasising the possibility of its use in this area as well. 16S/18S gene analysis with NGS usage allows for metagenomic analysis of any environmental sample and obtaining information on species diversity without the need for cloning or laborious electrophoretic separation. The versatility of the method is evidenced by the use of NGS technology to analyse the biodiversity of many environments, including water, air and soil [Be et al. 2015; Bell et al. 2013; Blais-Lecours et al. 2012; Delort and Amato 2018; Nonnenmann et al. 2010; Yergeau et al. 2012].

Among the most promising techniques for the assessment of microbiological differentiation of bioaerosol is also the droplet digital PCR (ddPCR). It was based on the fractionation of the test sample into droplets (formed in a water-oil emulsion) of nanolitres in volume each and an independent amplification of the DNA matrix in each of them. In the traditional qPCR technique, a single sample allows only one measurement. In the ddPCR, a split sample enables the measurement of thousands of independent amplification events in each drop, which is used, among other things, for the analysis of copy number variants, detection of rare sequences, quantification of microbial nucleic acids, analysis of gene expression and genotyping of single nucleotide polymorphism. Compared to other systems, the ddPCR technique uses a smaller volume of samples and reagents, which reduces the total cost of testing while maintaining the sensitivity and precision inherent to digital PCR [Hindson et al. 2011; Mazaika and Homsy 2014].

In order to precisely locate and recognise microbial aerosol particles and confirm their immunological reactivity, analytical protocols based on enzyme-linked immunosorbent assays (ELISA) and electrochemiluminescence (ECL) are also used as a mechanism to detect the binding reaction of specific epitopes' antibodies. These tests, based on poly- or monoclonal antibodies, are used, inter alia, in aeroallergens detection. However, these methods are still not widely used, mainly due to the very variable production of allergens in different environments, which is influenced by many factors, including the type of substrate, temperature and biodiversity of microorganisms. This variability makes it difficult to develop specific sensitive antibody-based immunoassays that would allow precise and unambiguous (without cross-reactivity) detection of the relevant bacterial or fungal allergens in a specific environment. However, these difficulties seem to pave the way for future innovative solutions in this area [IOM 2004; Schmechel et al. 2003, 2004].

## REFERENCES


Alvarez, A. J., M. P. Buttner, G. A. Toranzos et al. 1994. Use of solid-phase dfPCR for enhanced detection of airborne microorganisms. *Appl Environ Microbiol* 60(1):374–376.

Amann, R. I., W. Ludwig, and K. H. Schleifer. 1995. Phylogenetic identification and in situ detection of individual microbial cells without cultivation. *Microbiol Rev* 59(1):143–169.

Amato, P., M. Parazols, M. Sancelme, P. Laj, G. Mailhot, and A. M. Delort. 2007. Microorganisms isolated from the water phase of tropospheric clouds at the Puy de Dôme: Major groups and growth abilities at low temperatures. *FEMS Microbiol Ecol* 59(2):242–254.

Bakal, A., R. L. Górny, A. Ławniczek-Wałczyk, and M. Cyprowski. 2017. Molecular biology methods in assessing occupational exposure to harmful biological agents. *Podstawy i Metody Oceny Środowiska Pracy* 3(93):5–16.

Bartoszek, N., and M. Rosowski. 2017. Microscopic techniques in biological research. *Laboratorium* 9–10:12–21.

Be, N. A., J. B. Thissen, V. Y. Fofanov et al. 2015. Metagenomic analysis of the airborne environment in urban spaces. *Microb Ecol* 69(2):346–355.

Bell, T. H., E. Yergeau, C. Maynard, D. Juck, L. G. Whyte, and C. W. Greer. 2013. Predictable bacterial composition and hydrocarbon degradation in Arctic soils following diesel and nutrient disturbance. *ISME J* 7(6):1200–1210.

Benson, D. A., M. Cavanaugh, and K. Clark. 2013. GenBank. *Nucl Acid Res* 41(1):36–42.

Blais-Lecours, P., C. Duchaine, and M. Taillefer 2011. Immunogenic properties of archaeal species found in bioaerosols. *PLoS One* 6(8):e23326.

Blais-Lecours, P., P. Perrot, and C. Duchaine. 2015. Non-culturable bioaerosols in indoor settings: Impact on health and molecular approaches for detection. *Atmos Environ* 110(6):45–53.

Blais-Lecours, P., M. Veillette, D. Marsolais, and C. Duchaine. 2012. Characterization of bioaerosols from dairy barns: Reconstructing the puzzle of occupational respiratory diseases using molecular approaches. *App Environ Microbiol* 78(9):3242–3248.

Blais-Lecours, P., M. Veillette, D. Marsolais, Y. Cormier, S. Kirychuk, and C. Duchaine. 2014. Archaea des bioaérosols de fermes laitières, des poulaillers et des usines d'épuration des eaux usées: Leur rôle dans l'inflammation pulmonaire. Montreal: IRSST, R-827. https://www.irsst.qc.ca/media/documents/PubIRSST/R-827.pdf?v=2019-10-14. [accessed October 11, 2019].

Bosshard, P. P., S. Abels, R. Zbinden, E. C. Böttger, and M. Altwegg. 1996. Approaches for identification of microorganisms. *ASM News* 62:247–250.

Bosshard, P. P., R. Zbinden, S. Abels, B. Böddinghaus, M. Altwegg, and E. C. Böttger. 2006. 16S rRNA gene sequencing versus the API 20 NE system and the VITEK 2 ID-GNB card for identification of nonfermenting Gram-negative bacteria in the clinical laboratory. *J Clin Microbiol* 44(4):1359–1366.

Bridge, P., and B. Spooner. 2001. Soil fungi: Diversity and detection. *Plant Soil* 232(1/2): 147–154.

Brosseau, L. M., D. Vesley, N. Rice, M. N. Goodell, and P. Hairston. 2000. Differences in detected fluorescence among several bacterial species measured with a direct-reading particle sizer and fluorescence detector. *Aerosol Sci Technol* 32(6):545–558.

Chakravorty, S., D. Helb, M. Burday, N. Connell, and D. Alland. 2007. A detailed analysis of 16S ribosomal RNA gene segments for the diagnosis of pathogenic bacteria. *J Microbiol Meth* 69(2):330–339.

Chen, P. S., and C. S. Li. 2005. Sampling performance for bioaerosols by flow cytometry with fluorochrome. *Aerosol Sci Technol* 39(3):231–237.

Chen, P. S., and C. S. Li. 2007. Real-time monitoring for bioaerosols: Flow cytometry. *Analyst* 132(1):14–16.

Chi, M. C., and C. S. Li. 2007. Fluorochrome in monitoring atmospheric bioaerosols and correlations with meteorological factors and air pollutants. *Aerosol Sci Technol* 41(7):672–678.

Clarridge, J. E. 2004. Impact of 16S rRNA gene sequence analysis for identification of bacteria on clinical microbiology and infectious diseases. *Clin Microbiol Rev* 17(4):840–862.

Colwell, R. 2000. Viable but nonculturable bacteria: A survival strategy. *J Infect Chemother* 6(2):121–125.

Cox, C. S., and C. M. Wathes, eds. 1995. *Bioaerosols Handbook.* Boca Raton, FL: Lewis Publishers/CRC Press, Inc.

Crook, B. 1995a. Inertial samplers: Biological perspectives. In: *Bioaerosols Handbook*, eds. C. S. Cox, and C. M. Wathes, 247–267. Boca Raton, FL: CRC Press, Inc.

Crook, B. 1995b. Non-inertial samplers. In: *Bioaerosols Handbook*, eds. C. S. Cox, and C. M. Wathes, 269–284. Boca Raton, FL: CRC Press, Inc.

Delort, A. M., and P. Amato, eds. 2018. *Microbiology of Aerosols*. Hoboken: Wiley Blackwell, Hoboken.

Demirev, P. A., and C. Fenselau. 2008. Mass spectrometry for rapid characterization of microorganisms. *Annu Rev Anal Chem* 1:71–93.

Després, V. R., J. A. Huffman, S. M. Burrows et al. 2012. Primary biological aerosol particles in the atmosphere: A review. *Tellus B Chem Phys Meteorol* 64(1):15598.

Diel, R., A. Seidler, A. Nienhaus, S. Rüsch-Gerdes, and S. Niemann. 2005. Occupational risk of tuberculosis transmission in a low incidence area. *Respir Res* 14(6):35.

Directive 2000/54/EC of the European Parliament and of the Council of 18 September 2000 on the Protection of Workers from Risks Related to Exposure to Biological Agents at Work. *Official Journal of European Communities L*. 262/21, Brussels (with subsequent amendments: Commission Directive (EU) 2019/1833 of 24 October 2019 amending Annexes I, III, V and VI to Directive. 2000/54/EC of the European Parliament and of the Council as Regards Purely Technical Adjustments. *Official Journal of European Communities L*. 279/54).

Douwes, J., B. van der Sluis, G. Doekes et al. 1999. Fungal extracellular polysaccharides in house dust as a marker for exposure to fungi: Relations with culturable fungi, reported home dampness and respiratory symptoms. *J Allergy Clin Immunol* 103(3 Pt 1):494–500.

Duarte, S., C. Pascoal, A. Alves, A. Correia, and F. Cássio. 2010. Assessing the dynamic of microbial communities during leaf decomposition in a low-order stream by microscopic and molecular techniques. *Microbiol Res* 165(5):351–362.

Dutkiewicz, J., and L. Jabłoński. 1989. *Biologiczne szkodliwości zawodowe*. Warsaw: Państwowy Zakład Wydawnictw Lekarskich.

EN 13098. 2002. *Workplace Atmosphere: Guidelines for Measurement of Airborne Microorganisms and Endotoxin*. Warsaw: Polish Committee for Standardization.

EN 14031. 2002. *Workplace Atmospheres: Determination of Airborne Endotoxins*. Warsaw: Polish Committee for Standardization.

EN 14583. 2005. *Workplace Atmospheres: Volumetric Bioaerosol Sampling Devices: Requirements and Test Methods*. Warsaw: Polish Committee for Standardization.

Eng, J., R. M. Lynch, and R. S. Balaban. 1989. Nicotinamide adenine dinucleotide fluorescence spectroscopy and imaging of isolated cardiac myocytes. *Biophys J* 55(4):621–630.

Fisher, M. M., and E. W. Triplett. 1999. Automated approach for ribosomal intergenic spacer analysis of microbial diversity and its application to freshwater bacterial communities. *Appl Environ Microbiol* 65(10):4630–4636.

Francisco, D. E., R. A. Mah, and A. C. Rabin. 1973. Acridine orange-epifluorescence technique for counting bacteria in natural waters. *Trans Am Microsc Soc* 92(3):416–421.

Gabey, A. M., W. R. Stanley, M. W. Gallagher, and P. H. Kaye. 2011. The fluorescence properties of aerosol larger than 0.8 μm in urban and tropical rainforest locations. *Atmos Chem Phys* 11(11):5491–5504.

Gandolfi, I., V. Bertolini, R. Ambrosini, G. Bestetti, and A. Franzetti. 2013. Unravelling the bacterial diversity in the atmosphere. *Appl Microbiol Biotechnol* 97(11):4727–4736.

Gołofit-Szymczak, M., and R. L. Górny. 2010. Bacterial and fungal aerosols in air-conditioned office buildings in Warsaw, Poland: Preliminary results (winter season). *Int J Occup Saf Ergon* 16:407–418.

Górny, R. L., and A. Ławniczek-Wałczyk. 2012. Effect of two aerosolization methods on the release of fungal propagules from contaminated agar surface. *Ann Agric Environ Med* 19(2):279–284.

Griffin, D. W., D. L. Westphal, and M. A. Gray. 2006. Airborne microorganisms in the African desert dust corridor over the mid-Atlantic Ridge, Ocean Drilling Program, Leg 209. *Aerobiologia* 22(3):211–226.

Griffith, W. D., and G. A. L. DeCosemo. 1994. The assessment of bioaerosols: A critical review. *J Aerosol Sci* 25(8):1425–1458.

Harrison, D. E., and B. Chance. 1970. Fluorimetric technique for monitoring changes in level of reduced nicotinamide nucleotides in continuous cultures of microorganisms. *Appl Microbiol* 19(3):446–450.

Hairston, P. P., J. Ho, and F. R. Quant. 1997. Design of an instrument for real-time detection of bioaerosols using simultaneous measurement of particle aerodynamic size and intrinsic fluorescence. *J Aerosol Sci* 28(3):471–482.

Harrison, R. M., A. M. Jones, P. D. E. Biggins et al. 2005. Climate factors influencing bacterial count in background air samples. *Int J Biometeorol* 49(3):167–178.

Heidelberg, J. F., M. Shahamat, M. Levin et al. 1997. Effect of aerosolization on culturability and viability of gram-negative bacteria. *Appl Environ Microbiol* 63(9):3585–3588.

Henry, T., P. C. Iwen, and S. H. Hinrichs. 2000. Identification of Aspergillus species using internal transcribed spacer regions 1 and 2. *J Clin Microbiol* 38(4):1510–1515.

Hernandez, M., S. L. Miller, D. W. Landfear, and J. M. Macher. 1999. A combined fluorochrome method for quantitation of metabolically active and inactive airborne bacteria. *Aerosol Sci Technol* 30(2):145–160.

Hindson, B. J., K. D. Ness, D. A. Masquelier et al. 2011. High-throughput droplet digital PCR system for absolute quantitation of DNA copy number. *Anal Chem* 83(22):8604–8610.

Ho, J., and G. Fisher. 1993. *Detection of BW Agents: Flow Cytometry Measurement of Bacillus subtilis (BS) Spore Fluorescence*. Alberta, Canada: Defense Research Establishment Suffield, Medicine Hat, 1–34.

Hobbie, J. E., R. J. Daley, and S. Jasper. 1977. Use of Nuclepore filters for counting bacteria by fluorescence microscopy. *Appl Environ Microbiol* 33(5):1225–1228.

Hospodsky, D., N. Yamamoto, and J. Peccia. 2010. Accuracy, precision, and method detection limits of quantitative PCR for airborne bacteria and fungi. *Appl Environ Microbiol* 76(21):7004–7012.

Hung, L. L., J. D. Miller, and K. Dillon, eds. 2005. *Field Guide for the Determination of Biological Contaminants in Environmental Samples*. Fairfax, VA: AIHA.

IOM [Institute of Medicine]. 2004. *Damp Indoor Spaces and Health*. Washington, DC: National Academies Press.

Iwami, Y., S. Takahashi-Abbe, N. Takahashi, T. Yamada, N. Kano, and H. Mayanagi. 2001. The time-course of acid excretion, levels of fluorescence dependent on cellular nicotinamide adenine nucleotide and glycolytic intermediates of Streptococcus mutans cells exposed and not exposed to air in the presence of glucose and sorbitol. *Oral Microbiol Immunol* 16(1):34–39.

Jensen, P. A., B. Lighthart, A. J. Moohr, and B. T. Shaffer. 1994. Instrumentation used with microbial bioaerosols. In: *Atmospheric Microbial Aerosols: Theory and Applications*, eds. B. Lighthart, and A. J. Mohr, 226–284. New York: Chapman and Hall, Inc.

Jonsson, P., and F. Kullander. 2014. Bioaerosol detection with fluorescence spectrometry. In: *Bioaerosol Detection Technologies*, eds. P. Jonsson, G. Olofsson, and T. Tjarnhage, 111–141. New York: Springer-Verlag.

Jonsson, P., G. Olofsson, and T. Tjarnhage, eds. 2014. *Bioaerosol Detection Technologies*. New York: Springer-Verlag.

Karlsson, K., and P. Malmberg. 1989. Characterization of exposure to molds and actinomycetes in agricultural dusts by scanning electron microscopy, fluorescence microscopy and the culture methods. *Scand J Work Environ Health* 15(5):353–359.

Kaye, P. H., W. R. Stanley, E. Hirst, E. V. Foot, K. L. Baxter, and S. J. Barrington. 2005. Single particle multichannel bio-aerosol fluorescence sensor. *Opt Express* 13(10):3583–3593.

Kell, D. B., H. M. Ryder, A. S. Kaprelyants, and H. V. Westerhoff. 1991. Quantifying heterogeneity: Flow cytometry of bacterial cultures. *Antonie Leeuwenhoek* 60(3–4):145–158.

Kepner, R. L., and J. R. Pratt. 1994. Use of fluorochromes for direct enumeration of total bacteria in environmental samples: Past and present. *Microbiol Rev* 58(4):603–615.

Kermani, F., M. Shams-Ghahfarokhi, M. Gholami-Shabani, and M. Razzaghi-Abyaneh. 2016. Diversity, molecular phylogeny and fingerprint profiles of airborne Aspergillus species using random amplified polymorphic DNA. *World J Microbiol Biotechnol* 32(6):96.

Kethley, T. W., M. T. Gordon, and C. Orr. 1952. A thermal precipitator for aerobacteriology. *Science* 116(3014):368–369.

Kim, J. K., S. N. Jackson, and K. K. Murray. 2005. Matrix assisted laser desorption/ionization mass spectrometry of collected bioaerosol particles. *Rapid Commun Mass Spectrom* 19(12):1725–1729.

Kleefsman, W. A., M. A. Stowers, P. J. T. Verheijen, and J. C. M. Marijnissen. 2008. Single particle mass spectrometry: Bioaerosol analysis by MALDI MS. *Kona* 26:205–214.

Kox, L. F., J. van Leeuwen, S. Knijper, H. M. Jansen, and A. H. Kolk. 1995. PCR assay based on DNA coding for 16S rRNA for detection and identification of mycobacteria in clinical samples. *J Clin Microbiol* 33(12):3225–3233.

Kulkarni, P., P. A. Baron, and K. Willeke, eds. 2011. *Aerosol Measurement: Principles, Techniques, and Applications*. Hoboken: John Wiley & Sons, Inc.

Lange, J. L., P. S. Thorne, and N. Lynch. 1997. Application of flow cytometry and fluorescent in situ hybridization for assessment of exposures to airborne bacteria. *Appl Environ Microbiol* 63(4):1557–1563.

Ławniczek-Wałczyk, A., M. Gołofit-Szymczak, M. Cyprowski, A. Stobnicka, and R. L. Górny. 2017. Monitoring of bacterial pathogens at workplaces in power plant using biochemical and molecular methods. *Int Arch Occup Environ Health* 90(3):285–295.

Lee, S. H., H. J. Lee, S. J. Kim, H. M. Lee, H. Kang, and Y. P. Kim. 2010. Identification of airborne bacterial and fungal community structures in an urban area by T-RFLP analysis and quantitative real-time PCR. *Sci Total Environ* 408(6):1349–1357.

Leema, G., D. S. Chou, C. A. Jesudasan, P. Geraldine, and P. A. Thomas. 2011. Expression of genes of the aflatoxin biosynthetic pathway in Aspergillus flavus isolates from keratitis. *Mol Vis* 17(11):2889–2897.

Li, J. K., E. C. Asali, A. E. Humphrey, and J. J. Horvath. 1991. Monitoring cell concentration and activity by multiple excitation fluorometry. *Biotechnol Prog* 7(1):21–27.

Lighthart, B. 2000. Mini-review of the concentration variations found in the alfresco atmospheric bacterial populations. *Aerobiologia* 16(1):7–16.

Ma, Y., H. Zhang, Y. Du et al. 2015. The community distribution of bacteria and fungi on ancient wall paintings of the Mogao Grottoes. *Sci Rep* 5:7752.

Macher, J., ed. 1999. Bioaerosols: Assessment and control. *American Conference of Governmental Industrial Hygienists*, Cincinnati.

Madsen, A. M., A. Zervas, K. Tendal, and J. Lund Nielsen. 2015. Microbial diversity in bioaerosol samples causing ODTS compared to reference bioaerosol samples as measured using Illumina sequencing and MALDI-TOF. *Environ Res* 140:255–267.

Mainelis, G., A. Adhikari, K. Willeke, S. A. Lee, T. Reponen, and S. A. Grinshpun. 2002. Collection of airborne microorganisms by a new electrostatic precipitator. *J Aerosol Sci* 33(10):1417–1432.

Mainelis, G., S. A. Grinshpun, K. Willeke, T. Reponen, V. Ulevicius, and P. J. Hintz. 1999. Collection of airborne microorganisms by electrostatic precipitation. *Aerosol Sci Technol* 30(2):127–144.

Mainelis, G., K. Willeke, and P. Baron 2001. Electrical charges on airborne microorganisms. *J Aerosol Sci* 32(9):1087–1110.

Maron, P. A., D. P. H. Lejon, E. Carvalho et al. 2005. Assessing genetic structure and diversity of airborne bacterial communities by DNA fingerprinting and 16S rDNA clone library. *Atmos Environ* 39(20):3687–3695.

Mazaika, E., and J. Homsy. 2014. Digital droplet PCR: CNV analysis and other applications. *Curr Protoc Hum Genet* 82(7):24.1–7.24.13.

Miller, J. D., and J. C. Young. 1997. The use of ergosterol to measure exposure to fungal propagules in indoor air. *Am Ind Hyg Assoc J* 58(1):39–43.

Mukoda, T., L. A. Todd, and M. D. Sobsey. 1994. PCR and gene probes for detecting bioaerosols. *J Aerosol Sci* 25(8):1523–1532.

Muyzer, G., E. C. de Waal, and A. G. Uitterlinden. 1993. Profiling of complex microbial populations by denaturing gradient gel electrophoresis analysis of polymerase chain reaction-amplified genes coding for 16S rRNA. *Appl Environ Microbiol* 59(3):695–700.

Nadkarni, M. A., F. E. Martin, N. A. Jacques, and N. Hunter. 2002. Determination of bacterial load by real-time PCR using a broad-range (universal) probe and primers set. *Microbiology* 148(1):257–266.

Nehmé, B., Y. Gilbert, V. Létourneau et al. 2009. Culture-Independent characterization of archaeal biodiversity in swine confinement building bioaerosols. *Appl Environ Microbiol* 75(17):5445–5450.

Nehmé, B., V. Létourneau, R. J. Forster, M. Veillette, and C. Duchaine. 2008. Culture independent approach of the bacterial bioaerosol diversity in the standard swine confinement buildings, and assessment of the seasonal effect. *Environ Microbiol* 10(3):665–675.

Nilsson, H. O., J. Taneera, M. Castedal, E. Glatz, R. Olsson, and T. Wadström. 2000. Identification of Helicobacter pylori and other Helicobacter species by PCR, hybridization and partial DNA sequencing in human liver samples from patients with primary sclerosing cholangitis or primary biliary cirrhosis. *J Clin Microbiol* 38(3):1072–1076.

Nonnenmann, M. W., B. Bextine, S. E. Dowd, K. Gilmore, and J. L. Levin. 2010. Culture independent characterization of bacteria and fungi in a poultry bioaerosol using Pyrosequencing: A new approach. *J Occup Environ Hyg* 7(12):693–699.

Oppliger, A., N. Charriere, P. O. Droz, and T. Rinsoz. 2008. Exposure to bioaerosols in poultry houses at different stages of fattening; use of real-time PCR for airborne bacterial quantification. *Ann Occup Hyg* 52(5):405–412.

Orr, C., M. T. Gordon, and M. C. Kordecki. 1956. Thermal precipitation for sampling airborne microorganisms. *Appl Microbiol* 4(3):116–118.

Orsini, M., P. Laurenti, F. Boninti, D. Arzani, A. Lanni, and V. Romano-Spica. 2002. A molecular typing approach for evaluating bioaerosol exposure in wastewater treatment plant workers. *Water Res* 36(5):1375–1378.

Paez-Rubio, T., E. J. Viau, S. Romero-Hernandez, and J. Peccia. 2005. Source bioaerosol concentration and rRNA gene-based identification of microorganisms aerosolized at a flood irrigation wastewater reuse site. *Appl Environ Microbiol* 71(2):804–810.

Peccia, J., and M. Hernandez. 2006. Incorporating polymerase chain reaction-based identification, population characterization, and quantification of microorganisms into aerosol science: A review. *Atmos Environ* 40(21):3941–3961.

Perrott, P., G. Smith, Z. Ristovski, R. Harding, and M. Hargreaves. 2009. A nested real-time PCR assay has an increased sensitivity suitable for detection of viruses in aerosol studies. *J Appl Microbiol* 106(5):1438–1447.

Pillai, S. D., K. W. Widmer, S. E. Dowd, and S. C. Ricke. 1996. Occurrence of airborne bacteria and pathogen indicators during land application of sewage sludge. *Appl Environ Microbiol* 62(1):296–299.

Pöhlker, C., J. A. Huffman, and U. Pöschl. 2011. Autofluorescence of atmospheric bioaerosols: Fluorescent biomolecules and potential interferences. *Atmos Meas Tech Discuss* 4(5):5857–5933.

Pöschl, U. 2005. Atmospheric aerosols: Composition, transformation, climate and health effects. *Angew Chem Int Ed Engl* 44(46):7520–7540.

Prigione, V., G. Lingua, and V. F. Marchisio. 2004. Development and use of flow cytometry for detection of airborne fungi. *Appl Environ Microbiol* 70(3):1360–1365.

Prussin, A. J. II., L. C. Marr, and K. J. Bibby. 2014. Challenges of studying viral aerosol metagenomics and communities in comparison with bacterial and fungal aerosols. *FEMS Microbiol Lett* 357(1):1–9.

Ramachandran, D., R. Bhanumathi, and D. V. Singh. 2007. Multiplex PCR for detection of antibiotic resistance genes and the SXT element: Application in the characterization of Vibrio cholerae. *J Med Microbiol* 56(3):346–351.

Rappé, M. S., and S. J. Giovannoni. 2003. The uncultured microbial majority. *Annu Rev Microbiol* 57:369–394.

Reeslev, M., M. Miller, and K. F. Nielsen. 2003. Quantifying mold biomass on gypsum board: Comparison of ergosterol and beta-N-acetylhexosaminidase as mold biomass parameters. *Appl Environ Microbiol* 69(7):3996–3998.

Reponen, T., K. Willeke, S. Grinshpun, and A. Nevalainen. 1995. Biological particle sampling. In: *Bioaerosol Handbook*, eds. C. S. Cox, and C. M. Wathes, 751–778. Boca Raton, FL: CRC-Press.

Roszak, D. B., and R. R. Colwell. 1987. Survival strategies of bacteria in the natural environment. *Microbiol Rev* 51(3):365–379.

Rylander, R., M. Reeslev, and T. Hulander. 2010. Airborne enzyme measurements to detect indoor mould exposure. *J Environ Monit* 12(11):2161–2164.

Sanger, F., and A. R. Coulson. 1975. A rapid method for determining sequences in DNA by primed synthesis with DNA polymerase. *J Mol Biol* 94(3):441–448.

Schmechel, D., R. L. Górny, J. P. Simpson et al. 2004. The potentials and limitations of monoclonal antibody-based monitoring techniques for fungal bioaerosols. *Workshop on Methods of Bioaerosol Detection*, Karlsruhe, Germany, 8–9 July, 2004.

Schmechel, D., R. L. Górny, J. P. Simpson, T. Reponen, S. A. Grinshpun, and D. M. Lewis. 2003. Limitations of monoclonal antibodies for monitoring of fungal aerosols using Penicillium brevicompactum as a model fungus. *J Immunol Methods* 283(1–2):235–245.

Shahamat, M., M. Levin, I. Rahman et al. 1997. Evaluation of media for recovery of aerosolized bacteria. *Aerobiologia* 13(4):219–226.

Staley, J., and A. Konopka. 1985. Measurement of in situ activities of nonphotosynthetic microorganisms in aquatic and terrestrial habitats. *Annu Rev Microbiol* 39:321–346.

Stärk, K. D., J. Nicolet, and J. Frey. 1998. Detection of Mycoplasma hyopneumoniae by air sampling with a nested PCR assay. *Appl Environ Microbiol* 64(2):543–548.

Stewart, S., S. Grinshpun, K. Willeke, S. Terzieva, V. Ulevicius, and J. Donnelly. 1995. Effect of impact stress on microbial recovery on an agar surface. *Appl Environ Microbiol* 61(4):1232–1239.

Tan, M., F. Shen, M. Yao, and T. Zhu. 2011. Development of automated electrostatic sampler (AES) for bioaerosol detection. *Aerosol Sci Technol* 45(9):1154–1160.

Tanaka, D., Y. Terada, T. Nakashima, A. Sakatoku, and S. Nakamura. 2015. Seasonal variations in airborne bacterial community structures at a suburban site of central Japan over a 1-year time period using PCR-DGGE method. *Aerobiologia* 31(6):143–157.

Tong, Y., and B. Lighthart. 1999. Diurnal distribution of total and culturable atmospheric bacteria at a rural site. *Aerosol Sci Technol* 30(2):246–254.

Van Wuijckhuijse, A. L., M. A. Stowers, W. A. Kleefsman, B. L. M. van Baar, C. E. Kientz, and J. C. M. Marijnissen. 2005. Matrix-assisted laser desorption/ionisation aerosol time-of-flight mass spectrometry for the analysis of bioaerosols: Development of a fast detector for airborne biological pathogens. *J Aerosol Sci* 36(5–6):677–687.

Verreault, D., L. Gendron, and G. M. Rousseau 2011. Detection of airborne lactococcal bacteriophages in cheese manufacturing plants. *Appl Environ Microbiol* 77(2):491–497.

Wainwright, M., N. C. Wickramasinghe, J. V. Narlikar, P. Rajaratnam, and J. Perkins. 2004. Confirmation of the presence of viable but non-cultureable bacteria in the stratosphere. *Int J Astrobiol* 3(1):13–15.

Wang, C. C., G. C. Fang, and L. Y. Lee. 2007. Bioaerosols study in central Taiwan during summer season. *Toxicol Ind Health* 23(3):133–139.

Wang, C. C., G. C. Fang, and L. Y. Lee. 2008. The study of ambient air bioaerosols during summer daytime and nighttime periods in Taichung, Central Taiwan. *Environ Forensics* 9(1):6–14.

Wang, Z., T. Reponen, S. A. Grinshpun, R. L. Górny, and K. Willeke. 2001. Effect of sampling time and air humidity on the bioefficiency of filter samplers for bioaerosol collection. *J Aerosol Sci* 32(5):661–674.

WHO [World Health Organization]. 2009. *Guidelines for Indoor Air Quality: Dampness and Mould*. Copenhagen: WHO Regional Office for Europe.

Woese, C. R. 1987. Bacterial evolution. *Microbiol Rev* 51(2):221–271.

Yamamoto, S., A. Terai, K. Yuri, H. Kurazono, Y. Takeda, and O. Yoshida. 1995. Detection of urovirulence factors in Escherichia coli by multiplex polymerase chain reaction. *FEMS Immunol Med Mic* 12(2):85–90.

Yang, C. S., and P. Heinsohn. 2007. *Sampling and Analysis of Indoor Microorganisms*. Hoboken: John Wiley and Sons.

Yergeau, E., J. R. Lawrence, S. Sanschagrin, M. J. Waiser, D. R. Korber, and C. W. Greer. 2012. Next-generation sequencing of microbial communities in the Athabasca River and its tributaries in relation to oil sands mining activities. *Appl Environ Microbiol* 78(21):7626–7637.

# 7 Assessment of the Air and Surface Microbial Contamination Levels

*Agata Stobnicka-Kupiec*

## CONTENTS



## 7.1 ENVIRONMENTAL RISK

Biological agents are widespread in both occupational and non-occupational environments and exposure to bacteria, fungi or viruses in both of them is therefore common and often leads to adverse health effects in exposed individuals. Many of the biological agents are airborne and, when inhaled, may be responsible for different adverse reactions, ranging from simple irritations, through allergies and infections, to toxic reactions and development of numerous non-specific symptoms.

Any biological agent, the presence of which in the environment is undesirable, may be considered a contaminant. Although no environment, except for those specifically established e.g. in the pharmaceutical, biotechnology, health or scientific industries, is sterile and free from contamination, the presence of many biological agents at low concentrations may be considered 'normal'. However, the problem of contamination may arise when the level of such contamination increases above a certain limit that is considered acceptable for the given environment. Such a situation is often encountered in the case of microbiological contamination of the indoor environment. Harmful microbial agents constitute the most common contamination in the indoor environment as the components of bioaerosols. Being airborne, they can be easily inhaled; however, they may also enter the organism through the skin and mucous membranes [Górny 2004; Górny and Dutkiewicz 2002; Dutkiewicz and Jabłoński 1989; Macher 1999]. The health hazards resulting from such exposure are high and, contrary to popular belief, are not solely limited to pathogens. In the United Kingdom, it is estimated that around 200 people each year are diagnosed with adverse health effects caused by exposure to microbial agents, and the estimated social costs of treatment of such exposure (and resulting diseases) amount to ~£100 million per year [Górny 2010a].

A prerequisite for maintaining the proper condition of the indoor environment and complete health and comfort is precise control of the exposure and parameters influencing the level of environmental pollution caused by harmful microbial agents. It seems that on a global scale this problem could be at least partly solved by applying hygienic standards for environmental control of microbial agents.

## 7.2 HYGIENIC STANDARDS FOR ENVIRONMENTAL MICROORGANISMS

Hygiene standards, frequently understood in this context as threshold values, are legally established environmental limit concentrations for microbial agents harmful to health, acceptable from the exposure assessment point of view [PN-ISO 4225/Ak:1999]. Unlike most chemical and physical agents, there are no globally accepted criteria for assessing exposure to microbiological agents, or generally recognised normative values (reference/threshold limit values) and methodological recommendations for the quantitative and qualitative control of these agents [Macher 1999; Górny 2004; Górny et al. 2011]. Difficulties in formulating threshold limit values for environmental concentrations of microorganisms result from several factors. First and foremost, there is still a lack of satisfactory epidemiological data on the relationship between exposure to the specific agent and the health effects resulting from such exposure. This is mainly due to the fact that the sensitivity of each organism exposed to a given microbial agent is an individual feature of the organism, which in epidemiological terms translates into difficulties in unambiguous determination of the effects of such action. Moreover, despite the progress in development of sampling and analysis techniques, the global scientific database on the quantitative and qualitative characterisation of microbial agents is still insufficient and incomplete with regard to many environments. There is also a lack of standardization of measurement methods (e.g. there are no standard samplers) or experimental and analytical methodologies that could be widely applied in the analysis of harmful microbial agents. However, numerous initiatives are still being undertaken to address all these issues.

Due to the above-mentioned limited access to data describing the relationship between the concentrations of microbial harmful agents and the health effect caused by them, standards or their proposals, if any, are not generally applicable in practice. Most of the available standards or recommendations are based on the clinical picture of diseases caused by a given microbial agent and include the procedure for its collection and application of corrective and preventive measures without setting quantitative limits. Nevertheless, there are several numerical proposals for threshold limit values in the relevant literature, which help to interpret empirically obtained measurement data. They are usually either arbitrary or relative in nature. Table 7.1 summarises the most important features of both types.

The arbitrary prescriptive values determine the concentration levels of microbial agents (e.g. in relation to the whole microbiota or a specific species) that are considered acceptable or unacceptable. They are usually designated by individual

**TABLE 7.1**
**The Features of Numerical Reference Values for Harmful Microbiological Agents (HMA)**

| Numerical reference values | |
|---|---|
| **Arbitrary** | **Relative/comparative** |
| 1. Concentration of HMA, which is acceptable or unacceptable | 1. Usually based on simultaneous measurements of HMA in indoor and outdoor environment |
| 2. Usually for total microbial concentration, group of bioagents or given species | 2. If the indoor/outdoor ratio is <1, it is interpreted as "lack of contamination" or acceptable level of contamination |
| 3. Proposed by individual researchers, expert groups or as a result of cross-sectional/large-scale research studies | 3. Identification of indoor sources of HMA |
| 4. Not related to specific health outcome (does not reflect dose–response relationship) | 4. Suitable for qualitative or frequency of appearance comparisons |
| 5. Often equal or close to detection level of particular analytical method | |

[Górny et al. 2011]

researchers or groups of scientists (experts) or are based on the results of cross-sectional studies conducted in 'normal' environments, i.e. without reference to specific health effects caused by microbial agents (i.e. they do not specify the dose–effect relationship). Nevertheless, they determine the concentration levels that 'normally' occur in a given environment or in part of it, in which case any exceedance found by the measurement is treated as unusual and indicates the possibility of an additional source of pollution. Arbitrary values are also determined (usually at or near the detection limit of the method used) for microorganisms that cause serious adverse health effects.

In establishing relative standards values, the relationship between concentrations of microbial agents in specimens measured simultaneously in indoor and outdoor environments is usually used. As a rule, if the concentrations in the indoor environment are lower than those in the outdoor environment, then the hygienic quality of indoor environment shall be assessed as good or acceptable. The ratio of indoor and outdoor concentrations also indicates the possible existence of indoor emission sources. Relative assessment methods are also used in a qualitative comparison or when confronting the prevalence of e.g. specific genera or species of microorganisms [Górny 2004; Górny et al. 2011].

### 7.2.1 Strategies for Creating Hygienic Standards for Bioaerosols

An adequate strategy for the development of threshold limit values for bioaerosols shall include a research method and a number of environmental, source, quantitative and qualitative criteria which are of key importance for such a process. Figure 7.1

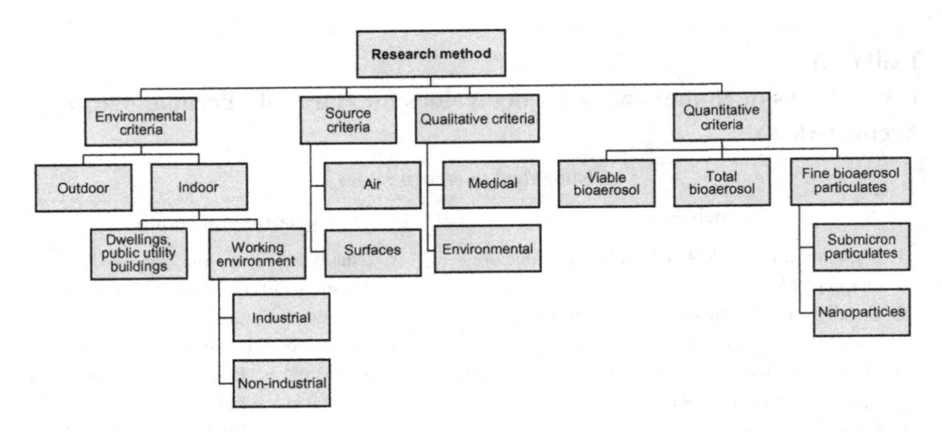


**FIGURE 7.1** Interdependence of key elements important for developing limit values for concentrations of microbiological agents [Górny et al. 2011].

illustrates their mutual relationships. The development of numerical limit values or their proposals shall be made both for the studied environments and for the types of environmental samples collected during measurements of microbial agents. The analysis of scientific literature indicates that some of the proposed threshold limit values relate to the outdoor environment (e.g. atmospheric air), and the rest to the indoor environment. For the indoor environment, both industrial and non-industrial facilities (e.g. offices, public utility buildings, schools, cinemas, shops, hospitals etc.) as well as residential environments (e.g. houses, flats or hotels) are included.

An integral part of each standard is its technical component, concerning the research method used to determine the numerical value of a standard. In short, traditional methods of collection and analysis of bacterial and fungal aerosols are based on the quantitative and qualitative evaluation of microbial vegetative cells, spores or conidia. Such measurement of airborne particles is usually relatively short and may not fully reflect the real air quality. Exposure assessment methods based solely on conventional (volumetric) air sampling, although substantively justified, may not reflect microbiological contamination reliably enough and may therefore only be considered an approximation of the real exposure level. Frequently, air sampling is not sufficient to identify problems with microbiological contamination or to establish a link between the contamination and the health effect it causes. This is partly because this type of measurement, which by its very nature is limited in time, and is often situated in a place a few to over a dozen metres from the emission source, does not always adequately represent the highest possible concentration of microbial agent being measured in the environment. Consequently, the assessment of the risks associated with exposure to bioaerosols should include, as a fixed part, identification of the microbial source and, if possible, the measurement of emission source strength of that source. Currently, however, (apart from a few prototypes) there are no commercially available instruments which would allow the performance of such types of measurements [Górny 2004, 2010a, 2010b; Górny et al. 2011].

Historically, the Koch's passive sedimentation method was used as a basis for the first proposals of acceptable concentrations of microorganisms in the environment

[Burge 1995; CEC 1993; Macher 1999; Reponen et al. 2001]. In accordance with contemporary requirements for microbiological testing of the air, it is necessary to use volumetric methods that enable active sampling of the air of a certain volume at a given time [Górny et al. 2011]. Nevertheless, in selected standards, especially those concerning the so-called 'special environments' (e.g. 'clean in operation' rooms), both (i.e. sedimentary and volumetric) methods of microbiological control of the air quality are still acceptable, with the addition of surface microbial quality control.

### 7.2.2 Proposals for Acceptable Concentrations of Harmful Microbiological Agents

As mentioned, despite the absence of generally applicable limit values for the concentration of microorganisms in indoor air, there are still attempts to develop them. A broad review of hygienic standards with respect to hazardous microbiological agents, i.e. bacteria, fungi and substances of microbial origin, in industrial and non-industrial settings, including special environments where high air purity is required, is discussed in detail (along with relevant comments) in publications by Brandys and Brandys [2007], Górny [2004] and Górny et al. [2011].

Against the background of world scientific literature, the Polish proposals for regulations in this area seem to be worth mentioning. As it is not possible to determine in a precise way a dose–effect relationship for almost all known harmful microbial agents, a reasonable alternative approach seems to be to set the threshold limit values for such agents based on the so-called 'environmental philosophy'. Based on this approach, it could be stated that multiple measurements of microbial agent concentrations in a given environment (or in its particular element) should enable the determination of what is 'typical and acceptable' and what is 'unusual and unacceptable'. As there are no widely acceptable legal acts formulating limit values for the concentrations of microbial agents in different environments, and with the undoubted social need for such regulations, wherever it is practically possible and, above all, scientifically justified, proposals for microbial threshold values are regularly being developed in Poland. At present, they are limited to the assessment of air pollution by microorganisms and bacterial endotoxins (Tables 7.2 and 7.3). These recommendations can be helpful not only in assessing exposure to microbial agents in the working, indoor or outdoor environment, but also in taking appropriate preventive measures. It is possible that if they are positively verified by science and practice, after several years, they could be transformed into widely used standards [Pośniak 2018; Górny 2010b; Górny et al. 2011].

At present, there is also a lack of generally established threshold values relating to the degree of surface contamination. In the scientific literature, there are proposals of standards, determining the hygienic quality of the surfaces of residential and public utility premises contaminated with fungal conidia (Table 7.4) [Kemp and Neumeister-Kemp 2010]. For buildings affected by microbial corrosion, the degree of surface contamination can also be determined using the D-A-N diagnostic scale – see Table 7.5 [Charkowska et al. 2005]. The use of the aforementioned scale makes it possible to determine the extent of mould biodeterioration of a building, including both air and surface contamination, based on measuring the concentrations of fungi and ergosterol (as a marker of their biomass).

**TABLE 7.2**
**Recommended Threshold Limit Values for Microorganisms and Endotoxins in Indoor Air Proposed by the Expert Group for Biological Agents**

| Microbial agents | Admissible concentration: Workstations contaminated with organic dust | Admissible concentration: Residential and public utility premises |
|---|---|---|
| Mesophilic bacteria | $1.0 \times 10^5$ CFU/m$^3$* | $5.0 \times 10^3$ CFU/m$^3$ |
| Gram-negative bacteria | $2.0 \times 10^4$ CFU/m$^3$* | $2.0 \times 10^2$ CFU/m$^3$ |
| Thermophilic actinomycetes | $2.0 \times 10^4$ CFU/m$^3$* | $2.0 \times 10^2$ CFU/m$^3$ |
| Fungi | $5.0 \times 10^4$ CFU/m$^3$* | $5.0 \times 10^3$ CFU/m$^3$ |
| Agents from the risk groups 3 and 4 | 0 CFU/m$^3$ | 0 CFU/m$^3$ |
| Bacterial endotoxins | 200 ng/m$^3$ (2000 EU/m$^3$)* | 5 ng/m$^3$ (50 EU/m$^3$) |

[Pośniak 2018]
*For respirable fraction the proposed values should be lowered by a half; EU – Endotoxin Units

**TABLE 7.3**
**Proposals for Assessment of Microbial Contamination of Outdoor Air**

| Microbial agents | Level of microbial contamination of outdoor air: Acceptable | Level of microbial contamination of outdoor air: Unacceptable |
|---|---|---|
| Bacteria (total number) | $\leq 5.0 \times 10^3$ CFU/m$^3$ | $>5.0 \times 10^3$ CFU/m$^3$ |
| Gram-negative bacteria | $\leq 2.0 \times 10^2$ CFU/m$^3$ | $>2.0 \times 10^2$ CFU/m$^3$ |
| Thermophilic actinomycetes | $\leq 2.0 \times 10^2$ CFU/m$^3$ | $>2.0 \times 10^2$ CFU/m$^3$ |
| Fungi | $\leq 5.0 \times 10^3$ CFU/m$^3$ | $>5.0 \times 10^3$ CFU/m$^3$ |
| Agents from the risk groups 3 and 4 | 0 CFU/m$^3$ | 0 CFU/m$^3$ |
| Bacterial endotoxins | $\leq 50$ EU/m$^3$ | $>50$ EU/m$^3$ |

[Górny 2010a; Pośniak 2018]
EU – Endotoxin Units

**TABLE 7.4**
**Mycological Surface Cleanliness Levels**

| Concentration of culturable fungal conidia on the surface | Hygienic assessment |
|---|---|
| $<0.5$ CFU/cm$^2$ | Low contamination |
| 0.5–1 CFU/cm$^2$ | Normal contamination |
| 1–2.5 CFU/cm$^2$ | Increased contamination |
| $>2.5$ CFU/cm$^2$ | Contaminated surface |
| $>12.5$ CFU/cm$^2$ | Extreme contamination |

[Kemp and Neumeister-Kemp 2010]

**TABLE 7.5**
**Diagnostic Scale of Mould Biodeterioration of Buildings – D-A-N**

| | | Indoor air | | Volume of building material | |
|---|---|---|---|---|---|
| Level of microbial contamination (D-A-N) | Description of the microbial contamination | Concentration of fungi, CFU/m³ | Concentration of ergosterol µg/m³ | Concentration of fungi, CFU/m³ | Concentration of ergosterol, µg/m³ |
| Acceptable | Normal state of contamination | <500 | <0.01 | <1000 | <2 |
| Emergency | Increased contamination, without active fungal development | 500–1000 | 0.01–0.03 | 1000–100000 | 2–4 |
| Dangerous | Active fungal development | >1000 | >0.03 | >100000 | >4 |

## REFERENCES


Brandys, R. C., and G. M. Brandys. 2007. *Worldwide Exposure Standards for Mold and Bacteria*. 7th ed. Hinsdale, IL: OEHCS, Inc., Publications Division.

Burge, H. 1995. *Bioaerosols*. Boca Raton, FL: Lewis Publishers/CRC Press, Inc.

CEC [Commission of the European Communities]. 1993. Biological particles in indoor environments: Report No. 12: Indoor Air Quality & Its Impact on Man. Brussels-Luxembourg.

Charkowska, A., M. Mijakowski, and J. Sowa. 2005. *Wilgoć, pleśnie i grzyby w budynkach*. Warsaw: Wyd. Verlag-Dashofer.

Dutkiewicz, J., and L. Jabłoński. 1989. *Biologiczne Szkodliwości Zawodowe*. Warsaw: PZWL.

Górny, R. L. 2004. Biologiczne czynniki szkodliwe: Normy, zalecenia i propozycje wartości dopuszczalnych. *Podstawy i Metody Oceny Środowiska pracy* 3(41):17–39.

Górny, R. L. 2010a. Aerozole biologiczne: Rola normatywów higienicznych w ochronie środowiska i zdrowia. *Medycyna Środowiskowa* 13:41–51.

Górny, R. L. 2010b. Normatywy higieniczne dla szkodliwych czynników mikrobiologicznych w ochronie powietrza wewnętrznego. *INSTAL* 4:38–45.

Górny, R. L., and J. Dutkiewicz. 2002. Bacterial and fungal aerosols in indoor environment in Central and Eastern European countries. *Ann Agric Environ Med* 9(1):17–23.

Górny, R. L., M. Cyprowski, A. Ławniczek-Wałczyk, M. Gołofit-Szymczak, and L. Zapór. 2011. Biohazards in the indoor environment: A role for threshold limit values in exposure assessment. In: *Management of Indoor Air Quality*, ed. M. R. Dudzińska, 1–20. Leiden: CRC Press/Balkema, Taylor & Francis Group.

Kemp, P., and H. Neumeister-Kemp. 2010. *Australian Mould Guideline*. Park, Australia: Osborne: The Enviro Trust.

Macher, J., ed. 1999. Bioaerosols: Assessment and control. *American Conference of Governmental Industrial Hygienists*, Cincinnati.

PN-ISO 4225/Ak:1999. Jakość powietrza. Zagadnienia ogólne. Terminologia (Arkusz krajowy). Warsaw: Polish Committee for Standardization.

Pośniak, M. 2018. *Czynniki szkodliwe w środowisku pracy: Wartości dopuszczalne*. Warsaw: CIOP-PIB.

Reponen, T., K. Willeke, S. Grinshpun, and A. Nevalainen. 2001. Biological particle sampling. In: *Aerosol Measurement: Principles, Techniques, and Applications*, ed. P. A. Baron, and K. Willeke, 751–777. New York: Wiley-Interscience.

# 8 Methods of Drying Buildings

*Rafał L. Górny*

## CONTENTS



## 8.1 WATER IN THE BUILDING

Water-damaged buildings pose serious technical, economic and health problems both for administrators and users. Almost all construction and finishing materials are characterised by a certain degree of porosity, permeability, susceptibility to swelling and other physiochemical changes caused by the impact of the surrounding environment [Badowska et al. 1974; Borusewicz 1985; Broniewski and Fiertak 1991; Karyś 2001, 2014; WHO 2009]. Structural elements of buildings, exposed to strong sunlight or low temperatures, undergo periodic changes in their spatial dimensions. That results in scratches and cracks as well as deformations in the form of warps and bulges, leading to reduction of material durability. Such changes foster penetration of moisture, resulting in numerous adverse effects. Above all, the cohesion of structural materials is weakened. Stone, brick, concrete, plaster, mortar that connects them or wood start to swell (they become more porous) and crumble (revealing the loss of substance), and thus their specific weight and mechanical properties decrease. These processes, in turn, result in creating internal stresses in a given material, which at the same time, with the impact of unevenly distributed external forces, lead to its deformation, loss of durability and, as a result, to destruction. Water also has a negative impact on the insulation properties of materials, because under its influence their coefficient of thermal conductivity increases, which results in higher heat loss in the building [Badowska et al. 1974; Bensted 2001; Borusewicz 1985; Broniewski and Fiertak 1991; Dylla 2009; Fijałkowski et al. 1987; Frössel 2007; Karyś 2001, 2014; WHO 2009].

In structural and finishing elements of a building, the presence of water usually results from:

- absorption of humidity from the air by porous and hygroscopic building materials and mortars,

- penetration of water vapour into pores and crevices and its condensation inside or on the surface of the structure,
- surface-wetting during natural precipitation or infiltration of water into walls from different sources,
- capillary action pulling water up from the ground.

Excessive humidity within the building envelope can also be caused by poor ventilation or its absence, inadequate insulation of the building, too tight windows and doors, and even the presence and activity of people in the building. For example, in a house inhabited by two people, the amount of water vapour that is produced during the day can condense into 5 litres of water. All these processes, however important, are usually stretched over time. In turn, massive water damage (technical breakdowns, inundations during fire, floods) cause the rapid insertion of huge amounts of water into the envelope of the building, which is absorbed by its individual structural and finishing elements. According to Karyś [2001, 2014], a brick wall of volume 1 $m^3$ is able to absorb about 300–350 litres of water. Flood studies show that mass humidity can reach in concrete wall 20%, in bricks 20–25%, in plastering mortar 10–15%, and in floors up to 10%. In the case of hygroscopic water absorption, the building material absorbs an amount of moisture proportional to its salinity. The destructive effect of water activity in such a case is more intense the more substances responsible for their chemical destruction permeate the structural materials.

As described above, from a technical point of view, it is clear that water has a negative impact on building materials, which lose their mechanical and insulating properties under its influence. Water, causing the chemical corrosion of materials, also usually opens the way for microbial growth. Nutrients, which are individual components of building materials, and the degree to which a given material is able to meet the requirements of a particular microorganism, in terms of the amount of moisture necessary to initiate its growth and maintain its subsequent development, cause the hygienic condition of a building to deteriorate drastically. This is manifested, among other things, by the increase in microbiological contamination of interiors caused by their bio-corrosion [Górny 2004]. Taking the above into account, the term 'dehumidification or drying of buildings' should be understood as a set of technical and technological activities causing a permanent decrease in the level of humidity in their structural elements (in practice up to 3–6% of the mass humidity), which enable the proper (i.e. mechanically and hygienically safe) exploitation of facilities. Therefore, if we want to effectively get rid of the undesirable presence of water from structural and finishing elements, we must first find and eliminate the cause of its appearance in the building, and then choose the optimal method to eliminate the adverse effects of the resulting humidity.

## 8.2 NATURAL DEHUMIDIFICATION

The headline process is lengthy and often requires from a few hundred days to even a dozen of years of enabling microclimatic conditions inside and in the surrounding of the dehumidified building. This method of dehumidification is usually applied to building partitions that have suffered from a failure of the water network or have been flooded, but have effective damp-proof isolation or waterproofing. The effectiveness of natural drying is conditioned by the construction and type of materials included in the partition and the speed of air flow at the dehumidified surface. The higher the dehumidification rate, the lower the relative humidity of the air surrounding a flooded or damp surface and the higher the temperature and speed of air movement in the surrounds of this surface. The efficiency of this process can be improved by installing blowers, fans or causing a draught near the dehumidified partition [Badowska et al. 1974; Karyś 2001, 2014; Kidd et al. 2010; Magott 2019; Rokiel and Magott 2013; Rokiel 2018].

The first stage of natural dehumidification is to drain water from damp surfaces sufficiently fast, which mainly depends on the difference in water vapour pressure in the material and in its surroundings. Usually this stage takes about one month. At this stage, the boundary of the humidity zone is being moved inward the partition. The next stage of drying is the elimination of the water remaining deep inside the wall. The efficiency of this process is affected by the diffusion resistance of the individual layers of the dehumidified structure and its geometric structure. According to Magott [2019], the approximate duration of natural dehumidification of a given partition can be expressed using the following formula:

$$t = a \times d^2$$

where: t – is the time of dehumidification of the wall to the level of equilibrium moisture (calculated in days); d – is the characteristic measurement of the partition equal to the largest distance on which the moisture must move from its interior to the surface (e.g. in case the partition dries on both sides, this measurement is equal to half of the wall thickness (calculated in cm); a – is a coefficient of moisture conductivity depending on the material properties and the degree of moisture (day/cm$^2$). Considering the above correlation, the drying time of 1.5 thick solid burnt brick wall is about 170 days, and of the same slag concrete wall about 680 days. It should also be noted that the effectiveness of natural dehumidification is closely linked to the weather conditions at a given time of year. Taking into account the fact that, e.g. in the summer in a temperate climate zone, there is a decrease in the wall moisture by about 1.5% per month, and in the autumn–winter period the process of natural dehumidification practically ceases, it can be assumed that the drying of the building partition two solid burnt bricks thick will be completed after about 1,000 days. Thus, the natural dehumidification method, although simple, is in practice only suitable for drying thin building partitions with a low degree of moisture [Karyś 2001, 2014; Magott 2019; Rokiel and Magott 2012; Rokiel 2018].

## 8.3 NON-INVASIVE ARTIFICIAL DEHUMIDIFICATION

Artificial dehumidification is a supporting and complementary action to natural dehumidification. It usually consists of increasing the temperature of the part of the building that is being dried and forcing air movement in its vicinity. The non-invasive methods include hot air, absorption, condensation, microwave and vacuum dehumidification [Badowska et al. 1974; Górny et al. 2013; Karyś 2001, 2014; Rokiel 2018].

*Hot air dehumidification* means the use of heaters that heat up the outlet air from the device to 50–250°C and force its circulation in the remedied room. The drying temperature in the room usually reaches between 35–37°C. Ensuring the drainage of moisture to the exterior of the dehumidified interior, e.g. by means of appropriate ventilation combined with simultaneous heating of the atmospheric air being inserted into the room (which is a necessary condition for carrying out the treatment efficiently) ensures the desired effectiveness of this process. When using hot air heating, both the dehumidified building partition's thickness and the method of finishing shall be taken into account. This method is most effective for drying thin building partitions with a low degree of moisture, preferably directly (usually up to three months) after they have been moistened or flooded. In this case, moisture is effectively removed from the surface of the dehumidified partition, and that which is being deposited in subsequent layers is forced inward and may return again (due to capillary forces and temperature gradient's changes) to the surface layers after the drying process is interrupted or not completed. When using hot air, it is also necessary to take into account the method of finishing of the dehumidified elements, as too high air temperature (e.g. above 80°C) may result in cracking of plaster layers [Karyś 2001, 2014; Kidd et al. 2010; Rokiel and Maggott 2012; Rokiel 2018].

*Absorption dehumidification* (Figure 8.1) consists of acquisition of water from flooded or moistened material through the surrounding air and bringing the material to the state of the so-called 'equilibrium moisture content/equilibrium relative humidity'. Dehumidification of the air that transports excess of moisture is done by its passing through a device with a water-absorbing agent, which may be for example

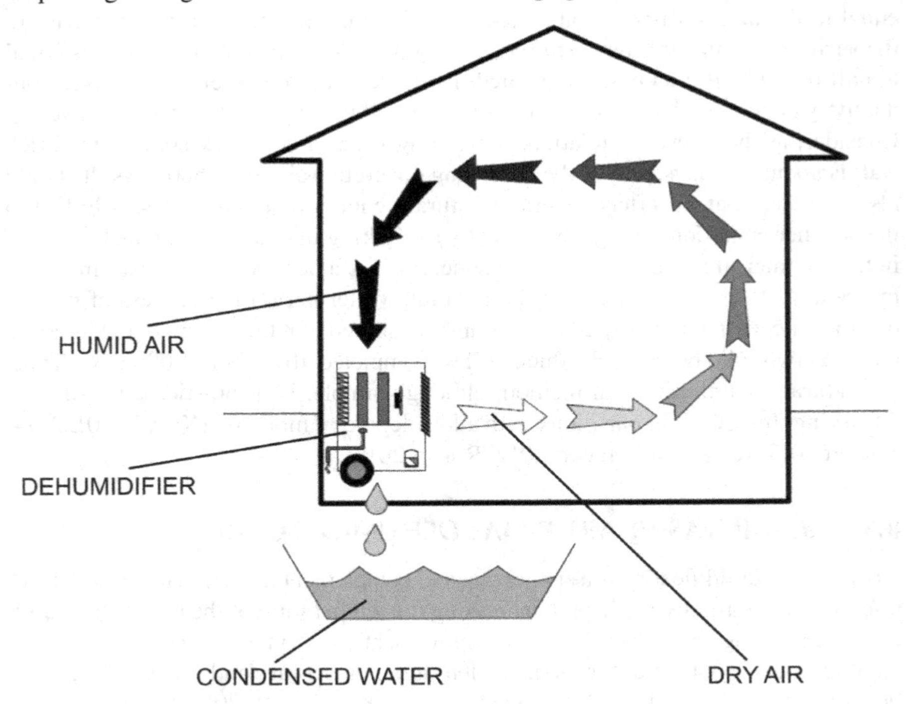


**FIGURE 8.1** Absorption drying.

a silicone gel, a silica gel or a lithium chloride. The air dried in this way is additionally heated and returns to the room, where it is saturated with water vapour again, and the moisture accumulated from the air is removed outside the (tightly closed during the whole process) interior's envelope. This method achieves the highest efficiency when the air humidity in the room is below 30%. The capacity of absorption air dehumidifiers used in this method is usually between 10–1000 litres of water per day [Broniewski and Fiertak 1991; Fijałkowski et al. 1987; Górny et al. 2013; Karyś 2001, 2014; Kidd et al. 2010; Rokiel and Magott 2012; Rokiel 2018].

*Condensation dehumidification* uses the phenomenon of condensation of water vapour contained in the air during contact with a material that has a lower temperature than the dew point temperature. Moistened air is drawn in through a device on which it evaporates, cools and liquefies. The heat gained this way from the humid air is transferred to the dehumidified interior, and the liquefied condensate (water) is discharged into the tank and removed. The drop in air humidity accelerates the evaporation and diffusion of moisture from the dehumidified building partitions. The parameters of the dehumidification process are selected so that within an hour the entire dehumidified air in a given room is circulated through the dehumidifier 3.5–4 times. Condensation dehumidifiers operate most efficiently in a temperature ranging from 20°C to 25°C and at relative air humidity from 30% to 90%. Depending on the power of the device, the drying efficiency can reach 1800 litres of water per day, and this parameter shows the highest values at high temperatures and high relative air humidity [Frössel 2007; Górny et al. 2013; Karyś 2014; Magott 2019; Rokiel and Magott 2012; Rokiel 2018].

*The microwave technique* used for drying various types of building partitions (walls, ceilings, floors etc.) consists of converting the energy of an electromagnetic field in the microwave radiation area of 2.5 MHz–300 GHz into thermal energy in the exhibited environment (Figure 8.2). This technique gives the possibility of freely shaping the size of the microwave area. Microwave devices, which are used

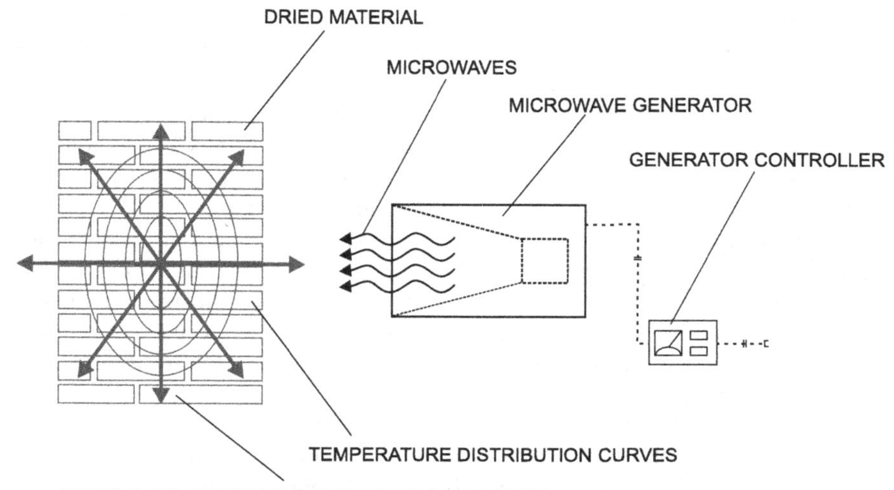


**FIGURE 8.2** Microwave drying.

for drying walls after flooding, inundations or insulation works blocking the migration of moisture, emit radiation of 2.5 MHz or 2.5 GHz and power up to several kilowatts. For safety reasons, the nature of this type of radiation requires directing the electromagnetic field and controlling the temperature of the dehumidified material (the safe level to which it is recommended to heat the dehumidified wall should not exceed 80°C). The advantages of microwave drying are: (a) speed – e.g. 1 $m^2$ of a 50 cm thick wall moistened by flooding can be dried in about four hours, and same wall moistened by absorbing water from the ground in about 30 hours; (b) effectiveness – it is possible to dehumidify walls up to 2.5 m thick; (c) non-invasiveness – microwave drying does not affect the structure of the dried material; (d) complexity – during dehumidification, viable organisms, including mould and bacteria, are destroyed; (e) lack of saltings on the plaster; (f) low costs – only a moistened or flooded surface is directly dehumidified, which to a large extent both speeds up the work and reduces the price of the treatment [Górny et al. 2013; Karyś 2001; Karyś 2014; Magott 2019; Rokiel and Magott 2013; Rokiel 2018].

The influence on the survival and cytotoxicity of microorganisms responsible for the bio-corrosion of construction and finishing materials has both thermal (related to the heat that is generated during friction of polar molecules, vibrating under the influence of the electromagnetic field) and microwave radiation effects (linked with selective absorption of radiation energy by the organism exposed to it, leading to the disintegration of covalent bonds of its nucleic acids). As the research has indicated [Górny 2013; Górny et al. 2007, 2013], in relation to contaminated building materials (inter alia, wallpaper, pine wood strips, plasterboard, gypsum plaster, solid brick, cement concrete or building stone, e.g. granite), the microwave effect is more important and is observed more often than the thermal effect. There is no single scheme for the selection of the most effective radiation parameters used to remove biological contamination from structural elements, although in a significant number of cases, for both xerophilic and hydrophilic microorganisms, it is effective to use radiation with a power density of 60 mW/$cm^2$ for 60 minutes. The purification of structural or finishing elements of a microbiologically contaminated building will therefore be effective if the microbiota responsible for the occurrence of bio-corrosion are first precisely characterised in terms of quality, and only after such 'diagnosis' is made the microwave purification treatment is carried out, using parameters of power density and radiation exposure time matching the species affinity of the dominant microorganisms. It should also be noted that some microorganisms, e.g. thermophilic actinomycetes (e.g. *Thermoactinomyces vulgaris*) or mould (e.g. *Penicillium brevicompactum*), under the influence of microwave radiation can become more expansive, and the reaction can manifest, for example, in the increase of cytotoxicity of their conidia. This phenomenon should be taken into account when combating bio-corrosion with the use of microwave technology is directly preceded by remedial actions carried out within the microbiologically contaminated building [Górny 2013; Górny et al. 2007, 2013].

The method of *magnetic interference* (also referred to as the magnetokinetic method) uses the Earth's natural magnetic field as an energy source. It is processed by a special device into waves of a strictly selected frequency and amplitude, influencing free ions in the capillaries of building partitions and causing transport of the water thus absorbed to the ground. The wall within the range of the field generated by the device is dried to the ground level around the building or lower to the down edge of the effective vertical insulation. The water moving back towards the ground removes significant (up to 30%) amounts of salt from the masonry, which makes drying the foundation, floors or plastered walls much more effective. After the dried partitions return to a naturally damp state, leaving the equipment creates a kind of horizontal insulation, which protects the walls from the rising of moisture from the ground again. Magnetokinetic devices are usually installed on the lowest floor of the building (usually in the basement) and do not require an electric power supply, but for their work to be effective, they must be installed in the building permanently (their effectiveness is usually checked every few years). With properly done vertical insulation, the magnetic interference method guarantees effective drying of the foundations. If vertical insulation in the building is lacking (or damaged), the dampness of the building envelope below ground level can be significantly reduced by this method. However, due to the physical aspect of the used phenomenon, this method cannot be applied when there is an accumulation of metal elements within the area of the building being dehumidified [Górny et al. 2013; Magott 2019; Rokiel 2018].

During the process of *vacuum dehumidification*, water from the damp material evaporates at low pressure. Under normal atmospheric pressure (1013 hPa) the water boils, turning into steam at 100°C, while when generating negative pressure of 100 hPa (90% vacuum) this value decreases to 45.8°C. An additional effect of creating a negative pressure is the increase of the pressure difference between water and vapour closed in the structure of the dried object and its surroundings. Both of these phenomena result in dehumidification at low temperature in a short time. The presence of heat, which by increasing internal stresses could damage the dried objects, is not a prerequisite for the efficient carrying out of the whole process. However, despite these advantages, the use of vacuum dryers is in practice limited to drying relatively small objects [Górny et al. 2013; Karyś 2001, 2014].

## 8.4 INVASIVE DEHUMIDIFICATION

In buildings that have been damaged by water, the adverse effects caused by the presence of water or the action of moisture can be reduced by making a membrane in the masonry by inserting an insulation layer, taking action to remove moisture continuously or by making a water-repellent or sealing barrier. Depending on the degree of moisture and the condition of the wall material, these methods can be used separately or combined [Broniewski and Fiertak 1991; Fijałkowski et al. 1987; Frössel 2007; Górny et al. 2013; Karyś 2001, 2014; Magott 2019].

*The insulation layer* can be made by manually or mechanically undercutting walls, underpinning strip foundations or mechanically pressing in the corrosion-resistant insulation steel sheet. The undercutting method is used for brick walls about 0.5 m thick (manual undercutting by carving) or 2 m (mechanical undercutting by saw, jet of liquid at 35 MPa or liquid with quartz sand). This activity is carried out around the building above the ground line up to 1 m long. After wedging the cut and preparing the substrate a suitable membrane is put on. It can be made of roofing tar paper, polyvinyl chloride film or epoxy resins. The space above the membrane

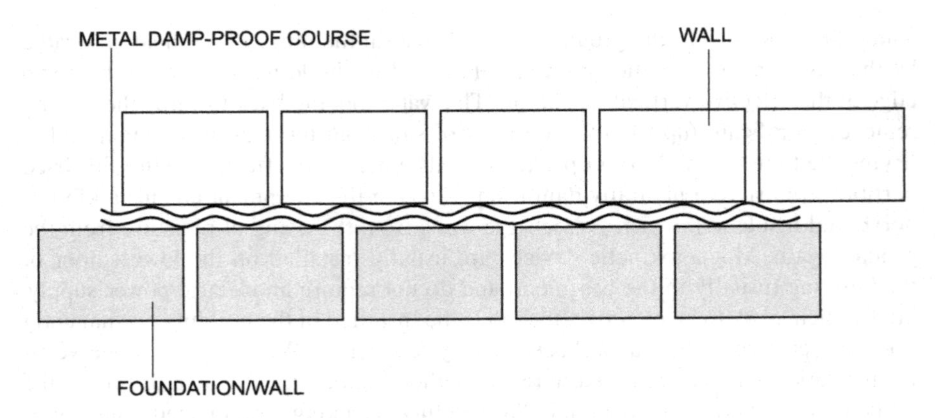


**FIGURE 8.3** Damp-proof course.

prepared this way is filled with cement mortar. *The underpinning technique*, on the other hand, consists of uncovering the strip foundation and making a several-centimetre-long support for the strip made of a material preventing capillary rise of ground water or other construction material (usually composite), if it is isolated from the strip with a proper insert. In the case of the *technique of mechanical or pneumatic membrane making*, a corrugated stainless steel sheet is inserted into the damp (usually brick) wall (usually perpendicular to the wall's face) in a percussive manner or by means of special presses, which mechanically insulates, limits or eliminates the transport of water through the structure of the building material (Figure 8.3) [Badowska et al. 1974; Górny et al. 2013; Karyś 2014; Kidd et al. 2010; Magott 2019].

Constant reduction of humidity in a building that has been damaged by water can be achieved by making: Knappen siphons, holes filled with a hygroscopic agent, ventilation panels, drainage ditches, perimeter drainage or using the electro-osmosis phenomenon. The methods of making holes in the wall are nowadays rarely used and were rather used in farm buildings. Holes up to 5 cm in the membrane were drilled upwards (*Knappen siphons*) or downwards, usually every 30–100 cm. In the first case, after the holes were drilled, heating spirals were inserted into them, which caused evaporation of water drained through the top of the holes, and this in turn resulted in the formation of saltings. In the second case, a strongly hygroscopic chemical compound was placed in the prepared holes (e.g. calcium or sodium chloride), and after its saturation it was replaced with a new one until the moisture was completely removed [Badowska et al. 1974; Górny et al. 2013; Karyś 2014; Kidd et al. 2010; Magott 2019; Rokiel 2018].

*Active ventilation panels* are used in buildings in which walls (usually load-bearing walls) are of considerable thickness. The panel can take the form of an internal or external structure and be made of ceramic brick, cavity brick or concrete blocks. The panel is placed at a distance of several centimetres from the damp wall and insulated from it with building paper or PVC foil, leaving intake vents in the lower part of the dehumidified room and establishing exhaust air vents in its upper part (under the ceiling, above the ground level). In this method, the excess moisture is transported to the outside of the building due to the movement of air in the gap created by the damp

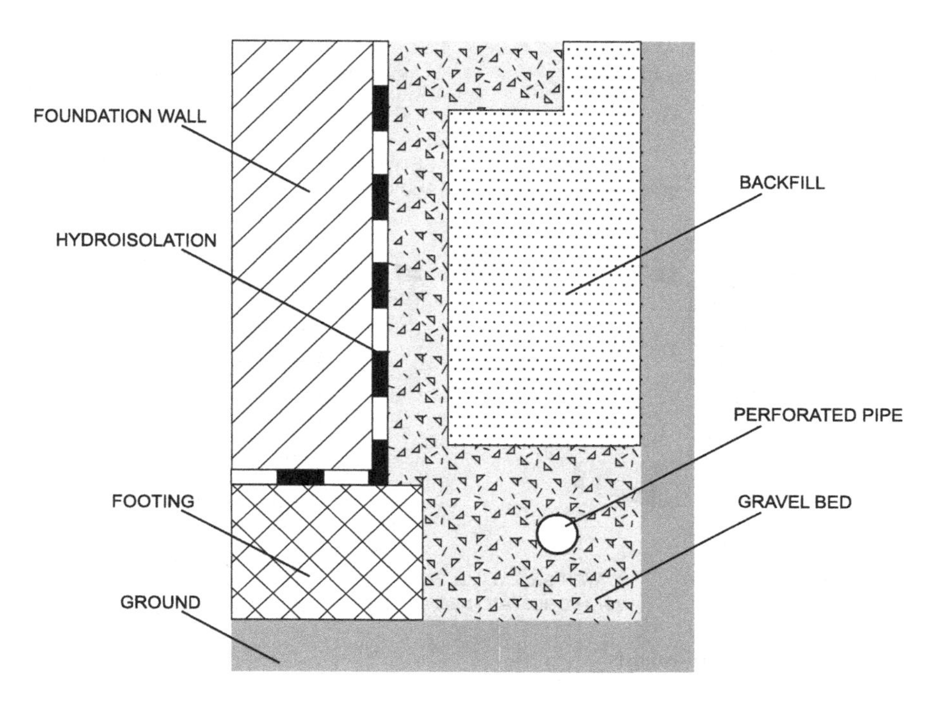


**FIGURE 8.4** Perimeter drain.

wall and the newly built panel. The construction of the panel should be preceded by plaster and mould removal from the fragment of the building envelope that is to be dehumidified [Górny et al. 2013; Karyś 2014; Kidd et al. 2010; Magott 2019].

*Making a French drain* overlapping the wall and draining the water away from the building envelope is a method often complementary to making an active external ventilation panel and protecting the previously damped wall from being flooded again, e.g. by rainwater. A much older and more commonly used way of protecting a building against this type of threat is *making a perimeter drain* (Figure 8.4). This method means placing, along the dried surfaces of the facade and above the level of the strip foundations, drains in the form of ceramic or PVC pipes draining the water away from the dehumidified structure. The consequence of making the drain is not only drainage of the land and protection against direct moisture, but also prevention of soil leaching [Badowska et al. 1974; Górny et al. 2013; Karyś 2014; Kidd et al. 2010; Magott 2019].

The reduction of the humidity in the building can also be achieved by *electro-osmosis technique*, which uses the phenomenon of potential difference between the lower and upper part of the brickwork dampened by water flow in its pores (Figure 8.5). The proper linking of electrodes to both these parts causes the flowing current (usually 24V ) to change the direction of moisture movement (i.e. it is directed towards the ground). This method of drying has a long-term effect and, depending on the moisture level and the thickness and type of the building material, can last up to several years. However, in this method, the coexisting phenomena of

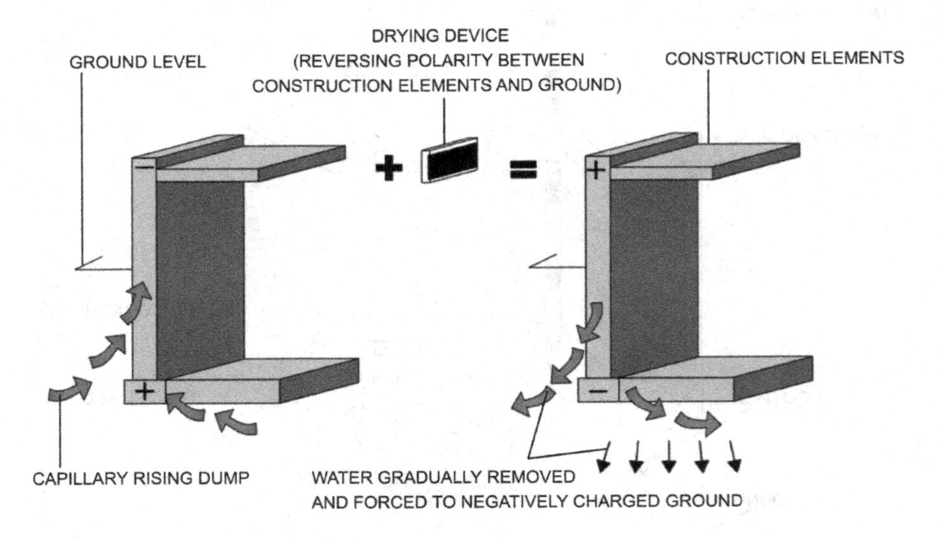


**FIGURE 8.5** Electro-osmosis.

electrode corrosion, current losses and the requirement of a frequent correction of the applied voltage and current intensity have resulted in no longer using this method today [Górny et al. 2013; Karyś 2014; Kidd et al. 2010; Magott 2019].

*The formation of a hydrophobic or impermeable membrane* is usually carried out by drilling into a damp wall a number of holes and inserting chemicals into them ('injections') to seal and close or hydrophobise pores of the wall. Among the methods that use such solutions, the following can be distinguished: gravity, low and high pressure, crystalline, electro- and thermo-injection. In the gravity method, chemicals (most often solutions of methyl silicone resin, silicates and methyl silicates, carbon-silicon polymer or siloxane emulsions, sometimes with the addition of a biocide) are inserted into the diagonally downwards directed (usually at an angle of 15–30°) holes in the wall, and their movement in the holes into the masonry is caused by gravity. In the low-pressure method, injections (in the form of alkaline silicates, methyl silicates, sodium or potassium aqueous glass, often combined with biocides) are applied mechanically to the previously prepared holes under low pressure (up to 1.5 MPa). The sealing and hydrophobic effect starts to be visible as soon as 24 hours after the treatment. In the high-pressure method, the injection rate (usually polyurethane foam, potassium water glass or epoxies) can be up to 10 MPa, so this technique should be used to protect walls of high mechanical strength against moisture. When the membrane is injected by the crystalline injection method, the active agent in the form of Portland cement mixed with sodium silicate, sodium phosphate or ethyl silicate acting as an activator, is inserted into the holes drilled in the wall (usually in one line in the case of horizontal insulation or in the form of a chessboard in the case of vertical insulation) and moistened with water, which prevents water from being

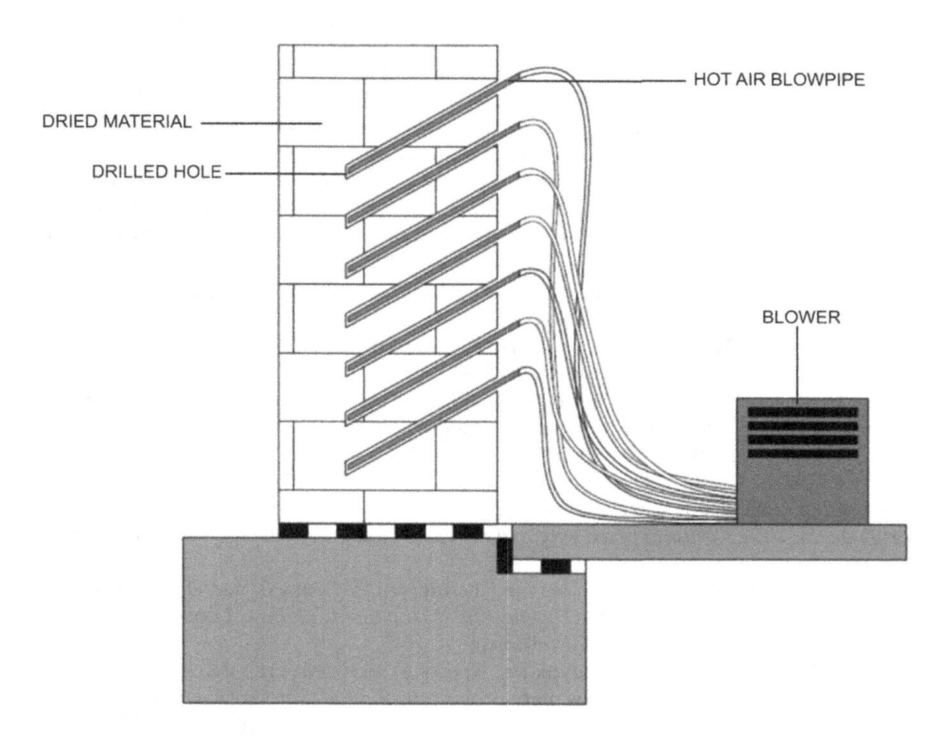


**FIGURE 8.6** Thermo-injection.

drawn up by the wall capillaries. The electro-injection technique, in turn, uses electrical voltage (24 V) to produce a potential difference which, by emptying the pores in the wall from the water, allows an active agent to be inserted into the previously prepared holes. In the thermo-injection method, heat is the factor emptying the pores and consequently allowing the insertion of an injection material (a solution of methyl silicone resin most often in isoparaffin, paraffin or bitumen) in them (Figure 8.6). Its source can be a thermoventilator, microwave device or electroresistance device. Under the influence of temperature (preliminary drying may take several days), the applied active agent reduces its viscosity, which increases its penetrating and insulating capacities, and effective hydrophobic blockade is obtained efficiently and quickly [Broniewski and Fiertak 1991; Fijałkowski et al. 1987; Górny et al. 2013; Karyś 2014; Kidd et al. 2010; Magott 2019].

## REFERENCES

Badowska, H., W. Danielecki, and M. Mączyński 1974. *Ochrona budowli przed korozją.* Warsaw: Arkady.

Bensted, J. 2001. The chemistry of efflorescence. *Cement Wapno Beton* 4:133–42.

Borusewicz, W. 1985. *Konserwacja zabytków budownictwa murowanego.* Warsaw: Arkady.

Broniewski, T., and M. Fiertak 1991. *Ochrona budowli przed korozją, T. I: Fizykochemiczne podstawy procesów korozyjnych w Budownictwie.* Kraków: Wydawnictwo Politechniki Krakowskiej.

Dylla, A. 2009. *Praktyczna fizyka cieplna budowli. Szkoła projektowania złączy budowlanych.* Bydgoszcz: Wydawnictwo Uczelniane UTP.

Fijałkowski, J., B. Ihnatowicz, and A. Kwiatkowski 1987. *Zabezpieczenia antykorozyjne w Budownictwie Przemysłowym: Poradnik projektanta.* Warsaw: Arkady.

Frössel, F. 2007. *Osuszanie murów i renowacja piwnic.* Warsaw: Polcen.

Górny, R. L. 2004. *Cząstki grzybów i bakterii jako składniki aerozolu pomieszczeń: Właściwości, mechanizmy emisji, detekcja.* Sosnowiec: Wydawnictwo IMPiZŚ.

Górny, R. L. 2013. Application of microwave radiation for removal of microbial contamination from building materials. *Ecology & Safety* 7(2):252–64.

Górny, R. L., A. Stobnicka, M. Gołofit-Szymczak, A. Ławniczek-Wałczyk, and M. Cyprowski 2013. *Usuwanie zanieczyszczeń mikrobiologicznych z budynków: Poradnik dla producentów i użytkowników urządzeń mikrofalowych oraz osób zatrudnionych w budownictwie.* Warsaw: Wydawnictwo CIOP–PIB.

Górny, R. L., G. Mainelis, A. Wlazło et al. 2007. Viability of fungal and actinomycetal spores after microwave radiation of building materials. *Annals of Agricultural & Environmental Medicine: AAEM* 14(2):313–24.

Karyś, J. 2001. Sposoby Osuszania Budynków. In *Ochrona budynków przed korozją biologiczną*, eds. J. Ważny, J. Karyś, 256–79. Warsaw: Arkady.

Karyś, J., ed. 2014. *Ochrona przed wilgocią i korozja biologiczna w budownictwie.* Warsaw: Grupa Medium.

Kidd, B., A. Tagg, M. Escarameia, B. von Christierson, J. Lamond, and D. Proverbs 2010. *Guidance and Standards for Drying Flood Damage Buildings.* London: Department for Communities and Local Government.

Magott, C. 2019. Osuszanie budynków: Sposoby osuszania zawilgoconych przegród budowlanych. https://www.muratorplus.pl/technika/hydroizolacje/proces-oraz-sposoby-osuszania-zawilgoconych-przegrod-budowlanych-aa-zxLg-2DAN-Mff2.html. [accessed September 10, 2019].

Rokiel, M. 2018. Naturalne i sztuczne metody osuszania budynków. *Forum Nowoczesnego Budownictwa* 6:16–24.

Rokiel, M., and C. Magott 2012. Osuszanie, Cz. 1. *Inżynier Budownictwa* 12:70–2.

Rokiel, M., and C. Magott 2013. Osuszanie, Cz. 2. *Inżynier Budownictwa* 1:105–6.

WHO [World Health Organization]. 2009. *Guidelines for Indoor Air Quality: Dampness and Mould.* Copenhagen: WHO Regional Office for Europe.

# 9 Removal of Microbial Corrosion from a Building

*Marcin Cyprowski*

## CONTENTS



Concerning the correct hygiene and sanitary conditions, the appropriate algorithm for removal of microbial corrosion in a building should include the following stages [Charkowska et al. 2005; Karyś 2014; Kemp and Neumeister-Kemp 2010]:

1) quantitative and qualitative identification of the microbiota within the building and in its vicinity,
2) determination of the extent and causes of microbial corrosion,
3) removal of causes of dampness and water damage,
4) designing construction and material solutions to prevent their re-wetting and biological contamination,
5) replacement of microbiologically damaged building structure elements with new ones,
6) removal of biological contamination from the elements that will remain in the building,
7) protection of new elements to be built in against biological corrosion.

Works related to the elimination of microbial corrosion should be applied to the entire facility or part of it, which constitutes a compact whole (for example, the entire floor or wing of a building). One should remember that a microbiologically infected building is dangerous, not only for people, but also for other neighbouring objects, as the spread of contaminants with the air and dust is the most common way of dissemination of this type of pollution in the environment [Charkowska et al. 2005].

## 9.1 METHODS FOR REMOVING MICROBIAL CONTAMINANTS

If microbiological hazards and moisture sources in the building have been identified, it must be decided how to remove the contamination from the corroded elements or entire contaminated structural or finishing elements from the building. Before taking appropriate action, it is important to consider who can perform such an activity. Available recommendations in this regard, coming from various institutions, are in principle consistent that in the case of small areas, i.e. not exceeding 1 $m^2$ of the microbiologically contaminated surfaces, cleaning operations can be carried out by the building users themselves. However, they should be particularly careful during such works and should wear personal protective equipment such as N95 type protective masks, rubber gloves and goggles. For the removal of medium-size microbiological contamination, i.e. from an area of 1–4 $m^2$, such works may still be carried out by the building users, but should be preceded by appropriate training, including safety procedures (e.g. regarding respiratory tract protection). However, in the case of massive contamination, any work can only be performed by trained and properly equipped professional services that will carry such work effectively and safely [NYCDH 2008; WHO 2009; CDPH 2016].

The choice of the most optimal way to clean contaminated surfaces should be based, apart from the very issue of effective disposal, on the necessity of minimising the emission of dust into the air containing the particles of microbial origin responsible for the contamination being removed. This will prevent secondary contamination of the building elements. The available methods for removing microbial contamination can be divided into two main groups: chemical and physical.

### 9.1.1 Chemical Methods of Microbial Contamination Removal

The use of chemical disinfectants (fungicides, antibacterial agents), despite their considerable effectiveness, is not generally recommended as a way of removing microbial contamination in buildings. There are two reasons for this position. Firstly, they can cause chemical pollution of indoor environment and, by that, create an additional threat for both the workers performing remediation and building users. Secondly, by killing viable microbial cells we do not eliminate all threats posed by them, as even after their death they may be responsible for adverse toxic or allergenic effects on humans [Górny 2004].

Nevertheless, the use of biocidal products is allowed in the case of microbial corrosion resulting from floodwaters or flooding of buildings with sewage [EPA 2008; NYCDH 2008]. All surfaces in flooded indoor areas must be cleaned and disinfected. Effective 'first-line' measures are usually those that contain chlorine. All damp surfaces (walls, floors, ceilings, stairs and other structural elements) have to be washed or wiped with them. For disinfection of buildings it is recommended to use: sodium hypochlorite, chloramine (in its 3% solution), chlorinated lime (dissolved in water in the ratio of 1:10) and quicklime (the so-called 'lime milk' as a 20% solution, mainly for disinfection of basements and livestock buildings). All disinfected sites should be washed again with warm water after one day from 'first-line' cleaning [EPA 2008; Karyś 2014].

### 9.1.2 Physical Methods of Microbial Contamination Removal

Physical removal of moulds or actinomycetes from buildings is the most often recommended way to remove microbial contamination [Kemp and Neumeister-Kemp 2010; Clements and Shum 2014]. Choosing the right cleaning strategy depends on the type of material that undergoes biodeterioration, therefore:

- in case of microbial contamination of porous materials, such as plasterboard, ceiling tiles, wallpaper, carpet/lino, insulation, it is recommended to remove them, as the available 'traditional' cleaning methods will not be efficient;
- when semi-porous materials, such as wood, are the subject of cleaning procedures, it is recommended to use an operation sequence consisting of scraping, scrubbing and vacuuming. However, it is recommended to check the integrity of the structure of contaminated material to decide whether it should be removed in the event of a massive contamination;
- bio-corrosion subjected to non-porous substrates, such as glass, plastics, concrete and metals can be cleaned by scrubbing, followed by appropriate washing.

The above activities can be carried out using 'dry' methods, e.g. efficient vacuum cleaners, protected by a HEPA filter, which eliminates particles larger than 0.3 μm (i.e. most fungi and bacteria). However, one should remember that a vacuum cleaner cannot remove all such contaminants, as they may remain in the pores of the material being cleaned. 'Wet' cleaning can be done by wiping the surfaces with appropriate detergents, vinegar solution or alcohol as well as by ultrasonic cleaning, steam-cleaning or high-pressure cleaning. Following this type of treatment, the surface to be cleaned must be thoroughly dried in order to prevent further microbial growth, and the material and equipment (cleaning cloths, sponges, brushes etc.) used for cleaning should be properly protected and removed.

Among the available methods of removing microbiological contamination in buildings, the use of microwave radiation (which enables both drying and removing microorganisms from building materials) is worth mentioning here. The use of this technique should be preceded by precise qualitative identification of the microbiota present on polluted surfaces and determination of the hydration state of identified microorganisms. Such information will allow adjusting the most effective parameters (the power and exposure time) of microwave radiation matching the species composition of the dominant microorganism in the treated building materials (see Chapter 8).

## 9.2 POST-REMEDIATION ACTIVITIES

Once all the activities aimed at removing or eliminating the microbial contamination and its causes have been carried out, it is necessary to perform a final verification in order to determine whether a given indoor area or building is already safe in terms of potential threats and free from such pollution. Amongst the measures verifying the effectiveness of the performed actions, the following steps should be included:

a) securing the building against renewed, uncontrolled access of water by controlling, inter alia, the surface and soil leaking of the building, the water intrusion within the building envelope, the humidity of the renovated indoor areas and their decontaminated equipment;
b) visual inspection confirming the absence of water damage, microbial contamination and odour nuisance, which may indicate an ongoing process of indoor biodeterioration;
c) microbial control of the air and surfaces of previously contaminated spaces;
d) microbial control of polluted building equipment surfaces, which will be placed within decontaminated spaces.

### 9.2.1 Visual Inspection

Visual inspection is the most important part of post-remediation activities and should focus on checking the following elements:

- whether all contaminated materials have been removed;
- whether the place is dry, clean and free of visible growth or mould odour and free of visible dust and other contaminants;
- whether there are no further signs of water damage and increased air and material humidity;
- whether people can re-occupy these spaces without fear of a recurrence of complaints connected with the presence of microorganisms.

### 9.2.2 Active and Passive Analytical Verification

Once remedial measures have been introduced, visible growth should no longer occur irrespective of the number or type of previously identified microorganisms. However, in the case of cleaning operations performed on large surfaces, quantitative analysis of microorganisms present in the air as well as on structural elements and building equipment should be carried out in order to verify the effectiveness of the performed work. It should be remembered that the area under examination will not be sterile even after a successful remediation process. Usually, as already shown in the study [Kleinheinz et al. 2006], a significant decrease in the number and types of microorganisms should be expected. Furthermore, their concentrations and taxonomic affiliation should be similar to those found in outdoor air, while concentrations of indicator microorganisms (i.e. *Aspergillus*, *Penicillium*, *Chaetomium*, *Stachybotrys*, *Streptomyces*) should be close to zero or at least significantly lower than in test samples taken from the polluted areas before remediation action. Such verification should always be carried out, especially in the case of premises occupied by vulnerable groups of inhabitants such as children, the elderly or sick people [EPA 2008].

Measurements of temperature and relative air humidity should be an essential and constant element of post-remediation control in buildings. The data obtained in this way allows the achievement of proper microclimate parameters in the rooms, preventing condensation, which promotes microbial growth. At relative air humidity below 60%, the development of filamentous microorganisms (moulds and some actinomycetes) is significantly restricted. The maintenance of proper microclimate parameters could be facilitated by the application of solutions elaborated for passive buildings, such as the introduction of a high-performance supply and exhaust ventilation system, which reduces heat losses within the building while guaranteeing efficient air exchange, which helps to reduce carbon dioxide concentrations. As recent studies show, high concentrations of this gas in building premises support mould growth [Koruba and Piotrowski 2017].

### 9.2.3 Checklist

When carrying out the process of remediation of microbial contamination, it is good to support oneself with a checklist, which will facilitate the performance of all necessary actions. An example of such a list follows:

A. Measures to assess the extent of water damage and bio-corrosion
- assess the size of the microbiologically contaminated area (in square metres);
- consider the possibility of the presence of contamination, which may not be visible during the initial inspection;
- remove/clean small contaminated areas and solve the problems related to moisture in a given indoor area before they become serious;
- select a person responsible for the coordination of medium- and high-level contamination removal activities;
- examine particularly carefully the areas that the building residents are paying attention to;
- identify the sources or causes of water damage or dampness;
- pay attention to the types of materials that have been contaminated or dampened (walls, carpets etc.);
- check the hygiene quality of the room and/or the building ventilation ducts and equipment;
- be in constant contact with specialists in water damage and bio-corrosion during the entire process of examination and evaluation.

B. Communication with building occupants at all stages of the remediation process
- appoint a person who will be answering questions from the building residents and providing explanations as to the reason and scope of the remedial measures being executed.

C. Remediation action plan
- adapt the action plan to suit the specific situation; seek professional help if necessary;

- plan the drying process for materials that have become damp but have not yet been bio-corroded to avoid contamination – this should be done within 48 hours from flooding;
- choose the method of cleaning the materials that have already been contaminated;
- choose personal protective equipment for workers performing remediation procedures;
- choose a way to limit the spread of microbial pollutants beyond the contaminated area;
- select people who are experienced and trained in such remediation actions to perform all necessary activities in a way that is safe for them and the surrounding area.

## 9.3 MATERIAL AND STRUCTURAL HAZARD PROPHYLAXIS

The best protection of buildings against microbial contamination is to shelter them from damp and lack of heating as well as ensuring proper ventilation, which will reduce the possibility of water vapour condensation on structural and finishing elements. In this respect, it is important to introduce appropriate rules for the use of specific indoor areas during everyday activities such as cooking, bathing or laundry. In buildings equipped with ventilation or air- conditioning systems, it is advisable to clean it on a regular basis, replacing during the maintenance procedure the air filters, which will reduce the penetration of microorganisms along with atmospheric air into the building. It is also necessary to perform the specific renovation works, which limit the penetration of moisture into the building.

## 9.4 METHODS OF CONSTRUCTION MATERIAL PROTECTION AGAINST MICROBIAL CORROSION

The following methods can be used to effectively protect water-damaged building materials against the reappearance of microbial corrosion [Karyś 2014]:

- lubrication (used to remove moulds from masonry and wood in the case of shallow infestation); it consists of covering (several times) the contaminated surface with a biocidal product;
- spraying (used to remove bio-corrosion in hard-to-reach areas such as deep cracks or narrow crevices); the biocidal product in aerosol form is applied several times to contaminated surfaces;
- bathing (consisting in immersing the entire element being disinfected in a biocidal liquid); depending on the agent used and the structure of the element being disinfected, the bath may last up to several hours;
- hot air heating (the action of hot air at 50–60°C for 1–3 days ensures effective removal of microbiological causes of corrosion, but may adversely affect structural elements of the building, especially wood, gypsum or paper); it is a self-contained or complementary disinfection treatment;

- dry mould removal (carried out using powder preparations which, when dissolved on wet or damp surfaces, diffuses into contaminated materials); this method is usually used to remove microbiological contamination from wooden materials;
- gassing (a fast and highly effective method of decontamination); due to the agents used, it requires hermetisation of indoor areas during treatments;
- drilling of holes with subsequent introduction of the biocide in liquid or semi-solid form (biocidal pastes or liquids are introduced in a sufficient quantity into the holes drilled in the contaminated material); diffusion of the active agent neutralises the microbiological agents; this method is usually used for fungicidal treatment of hard-to-reach spaces;
- firing (a form of thermal removal of microbiological contamination); due to the drastic nature of this method (it uses gas or petrol burners), firing is used to remove contamination from permanent masonry or this type of structural elements.

## REFERENCES

California Department of Public Health. 2016. Mold or moisture in my home: What do I do? https://www.cdph.ca.gov/Programs/CCDPHP/DEODC/EHIB/CPE/CDPH%20Document%20Library/Mold/MMIMH_English.pdf [accessed October 11, 2019].

Charkowska, A., M. Mijakowski, and J. Sowa. 2005. *Wilgoć, pleśnie i grzyby w budynkach*. Warsaw: Wydawnictwo Verlag-Dashofer.

Clements, L., and M. Shum. 2014. *Mould Remediation Recommendations*. Vancouver: National Collaborating Centre for Environmental Health at the British Columbia Centre for Disease Control.

DOHMH. [New York City Department of Health & Mental Hygiene]. 2008. *Guidelines on Assessment and Remediation of Fungi in Indoor Environments*. New York, NY: New York City Department of Health & Mental Hygiene. https://www1.nyc.gov/assets/doh/downloads/pdf/epi/epi-mold-guidelines.pdf. [accessed October 11, 2019].

EPA [United States Environmental Protection Agency]. 2008. *Mold Remediation and Schools and Commercial Buildings. Office of Air and Radiation, Indoor Environments Division (6609-J) EPA 402-K-01-001*. Washington, DC: United States Government Printing Office.

Górny, R. L. 2004. Filamentous microorganisms and their fragments in indoor air: A review. *Ann Agric Environ Med* 11(2):185–97.

Karyś, J., ed. 2014. *Ochrona przed wilgocią i korozją biologiczną w budownictwie*. Warsaw: Grupa Medium.

Kemp, P., and H. Neumeister-Kemp. 2010. *Australian Mould Guideline*. Park, Australia: Osborne: The Enviro Trust.

Kleinheinz, G. T., B. M. Langolf, and E. Englebert. 2006. Characterization of airborne fungal levels after mold remediation. *Microbiol Res* 161(4):367–76.

Koruba, D., and J. Z. Piotrowski. 2017. Analiza wpływu koncentracji ditlenku węgla, temperatury i wilgotności względnej powietrza na liczebność jednostek mogących tworzyć kolonie grzybów pleśniowych. *Fizyka Budowli w Teorii i Praktyce* 9(3):13–8.

WHO [World Health Organization]. 2009. *Damp and Mould: Health Risks, Prevention and Remedial Actions: Information Brochure*. Copenhagen: WHO Regional Office for Europe.

# 10 Microbiological Contamination of Indoor Environments in Legal Terms

*Anna Ławniczek-Wałczyk*

**CONTENTS**



Because of the lack of widely accepted methods for the assessment of exposure to harmful microorganisms in the indoor environment, the control of parameters limiting their growth is of key importance. Among them, the control of moisture levels plays an essential role in reducing microbiological corrosion of buildings. The easiest way to achieve this goal is to design, maintain and improve buildings in accordance with the Building Code, including the laws of building physics. Microbial growth in indoor environments can be prevented or minimised by regular maintenance work aimed at eliminating the causes of moisture and water intrusions and visible signs of bio-corrosion, as well as removing construction/finishing materials damaged by water and microorganisms. The exposure of building users to microorganisms and the resulting adverse health effects can already be prevented at the level of building design. The use of appropriate construction and finishing materials, resistant to microbial colonisation, and planning of construction works or remedial actions in consultation with building mycology specialists will undoubtedly enable keeping the building in good condition. Due to the fact that microbiologically contaminated indoor spaces are a significant public health problem, it becomes important to raise the general awareness of building designers, owners and users about health risks associated with the bio-corrosion of building materials. This type of informative measures should first of all be addressed to the people from the so-called 'high risk' groups, i.e. those suffering from asthma, lung diseases or diseases of allergic origin as well as parents and their children, elderly people and their carers, and people inhabiting council flats, which due to general housing conditions (usually dark and poorly ventilated) as well as low renovation investments are often characterised by elevated moisture levels. Information activities in this area should result in appropriate steps being taken both to improve indoor hygiene and to change users' behaviour in order to improve their quality of life.

## 10.1 LEGAL ACTS

State and local authorities usually have the legal ability to directly and indirectly change the hygiene quality of buildings damaged by bio-corrosion. Direct measures usually include changes in general regulations and the Building Code, whereas indirect measures focus, e.g. on conducting health programmes or information campaigns. In numerous countries, building law is regulated at national and local level, and specific requirements for buildings may differ, inter alia, in terms of their climate adaptation (usually related to ventilation, heating, drainage or resistance of structures to external weather conditions). A well-made building project by an architect or building designer must take into account at the beginning of the design process the possible influence of external and internal factors on building materials. In the European Union countries, in accordance with the requirements of Regulation (EU) No. 305/2011 of the European Parliament and of the Council of 9 March 2011 laying down harmonised conditions for the marketing of construction products, the construction products, and works should be made in such a way that they do not endanger health and hygiene, both for users and the environment. Also, in accordance with Directive 2010/31/EU of the European Parliament and of the Council of 19 May 2010 on the energy performance of buildings, which promotes energy-efficient construction and establishes new requirements for technical installations in buildings, in order to reduce heat loss, it is recommended to reduce the occurrence of the so-called 'thermal bridges' during the design phase of a building or during its upgrading [Directive 2010/31/EU]. This is to prevent the freezing and dampening of building envelopes, and, consequently, bio-corrosion of the building. Moreover, each country of the European Union has its own legal regulations in this area. For example, in Poland, the general technical requirements of buildings are regulated by the provisions of the Construction Law [Journal of Laws of 1994, no. 89, item 414], while the Regulation of the Minister of Transport, Construction and Maritime Economy of 5 July 2013 [Rozporządzenie Ministra Transportu, Budownictwa i Gospodarki Morskiej, Dz. U. z 2013 r., poz. 926] addresses in more detail the issue of the undesirable presence of water within the building envelope and related development of mould.

In slightly broader, global perspective, it can be noticed that probably the most changes, whether in the building code or in other local regulations on the protection of buildings against bio-corrosion, have been introduced in the US. For instance, in 2007 in the state of Maine, regulations were introduced into the building code to reduce moisture and mould growth in buildings. They concerned, among others, the methods of installing windows and doors, foundation and basement insulations as well as placing vapour diffusion retarders on the warm (winter) side of insulation surfaces (e.g. walls, ceilings or floors). Other recommendations introduced in Maine include those relating to the education of government officials and social campaigns on the harmful effects of moulds and to the protection of workers employed to remove it. A certification system has also been introduced for specialists and companies providing expertise and performing mould removal and remediation works [MEDEP 2007].

In 2010 in New York City, the NYC Green Codes Task Force proposed over 100 amendments to the Construction Law. Their recommendations concerned, *inter alia*, the use of building materials (gypsum board or cement board) that are resistant to moisture and moulds and the introduction of training for officials in charge of enforcing the regulations in order to identify the causes of mould growth in buildings. They also addressed the issue of providing guidance on recommended methods for the hygiene assessment of facilities and the types of remedial actions available, as well as the provision of information material to vulnerable groups of people and construction contractors and designers. At the same time, it was recommended that research should be conducted both on the use of modern building and finishing materials in the civil engineering, showing resistance to bio-corrosion and on the adverse health effects associated with staying in a 'sick building' [Urban Green Building Council 2010]. From 2015, the State of New York requires a special licence from the State Department of Labor for companies dealing with mould damage of buildings. The ultimate goal of such action is both to protect clients ordering such construction work from unsuitable companies and to ensure that remedial work is conducted in accordance with the applicable guidelines [New York State Labor Law 2015].

Similar solutions were also introduced in 2012 in New Hampshire. In 2015, the State of California adopted a law stating that mould is a factor that lowers housing conditions and allows enforcement from property owners to deal with the problem of microbiological contamination and remove the causes of its formation to stop the excessive spread of pollution within the building. The regulations forced the owner to immediately remove visible signs of mould contamination and inspect such interiors, and were also introduced in 2015 in Virginia. Furthermore, in the states of Montana, New Jersey, Ohio, Virginia and Washington, the seller or landlord of the property with visible marks of moisture and mould must inform the buyer or tenant of this fact and such action is required by law [Major and Boese 2017].

## 10.2 STANDARDS

The amount of scientific evidence on the relation between the bio-corrosion of buildings and the worsening of the health of their occupants has increased significantly over the last 30 years. During this period, numerous local, national and even international programmes have been developed which summarise the available knowledge in this field and promote actions to improve the air quality in industrial, public and non-industrial environments. The most important studies in this area include the Institute of Medicine 2004 report entitled 'Damp Indoor Spaces and Health' [IOM 2004] and the World Health Organization's 2009 report entitled 'Dampness and mould' [WHO 2009]. Both documents emphasise that people living in a damp indoor environment experience numerous health problems of an allergic, inflammatory, toxic and infectious nature. Both reports state that in order to avoid such adverse health effects, it is essential to take measures to reduce or eliminate moisture and microbiological contamination from internal surfaces and construction and finishing elements of the building. Both documents recommend monitoring air humidity levels as well as inspecting and quickly and accurately removing water-damaged and microbiologically contaminated construction and finishing elements. The WHO also recommends that, where possible and available, building regulations or requirements should be adapted to existing local climatic conditions. This will enable effective monitoring of microbial growth through the control of moisture levels in buildings and strive to achieve the desired indoor air quality.

The recommendations of IOM and WHO also highlight the need to undertake remediation actions in premises of poor hygienic quality, whose users face dampness and bio-corrosion. The above-mentioned documents also emphasise that both residents and workers must be protected when performing remediation work related to the removal of microbiological corrosion. They also recommend the control of adverse health effects provoked by microbial agents on people in the buildings affected by bio-corrosion. They also state that, in many cases, building codes and regulations relating to the use of buildings should be modified to take better care of indoor air quality.

One of the elements of protection of a building and its users against the effects of bio-corrosion is the use of construction and finishing materials resistant to microorganisms and water in its various forms [Klemm 2010]. According to the aforementioned Regulation No. 305/2011 of the European Parliament and of the Council, in the EU Member States [2011], all construction products must comply with certain performance requirements in order not to present a health and safety risk for both users and the environment. With regard to construction products, there are also a number of standards setting down the requirements for products to be resistant to bio-corrosion and chemicals to be used for the disinfection of microbiologically contaminated products. An example of this is the European standards: EN 350:2016-10 [2016], which provides information on methods for determining and classifying the durability of wood and wood-based materials against wood-destroying agents, including fungi and bacteria; EN 335:2013-07 [2013], which discusses the durability of wood and wood-based materials under exposure to biological agents; or EN206+A1:2016-12 of 2016, which provides the basic requirements for the concrete lagging for corrosion protection [2016].

The methods of remediation work are also widely discussed by many guidelines, which are often adapted to local climate conditions. They are aimed at professionals as well as building owners and users, who are less familiar with the microbial corrosion problems. In 2015, a team consisting of experts from the US Environmental Protection Agency (EPA) developed guidelines on mould removal after flooding and other water-related disasters. These guidelines, among other things, discussed the use of biocides (and the associated risks) for the disinfection of microbiologically contaminated surfaces and provided information on the health effects of exposure to moulds, methods of protection against microorganisms during remediation works, and ways of safely removing microbiologically contaminated elements from affected indoor spaces (houses, public buildings or schools) [EPA 2008; Brennan et al. 2015].

Key information on the risks of mould removal from water-damaged houses was also summarised in the guides of the US National Center for Healthy Housing (2008) and the New York City Department of Health & Mental Hygiene (2008). Both of them include practical work safety tips for activities during mould cleanup, removal and remediation processes in buildings.

The performance of remediation work related to the removal of water damage and signs of bio-corrosion results in exposure of workers to harmful microbiological agents. Means of collective protection and personal protective equipment should always be adapted to the type of hazard and the ways in which harmful agents may be spread in the environment. Requirements for safe work, when exposed to fungi and bacteria in water-damaged and microbiologically contaminated buildings, are discussed in several legal acts and guidelines. Across the ocean, the results of expert work in this area are included in the Occupational Safety and Health Administration (OSHA) guidelines of 2003 [OSHA 2003] as well as in recommendations of the EPA expert team, Department of Housing and Urban Development (HUD), National Institutes of Health (NIH) and OSHA [EPA, HUD, NIH and OSHA 2015]. In turn, in the European Union the main document related to the protection of workers is Directive 2000/54/EC on the protection of workers from risks related to exposure to biological agents at work [Directive 2000/54/EC]. It specifies the basic obligations of the employer to assess and document the occupational risks associated with the work, to take the necessary preventive measures to reduce the risks, to inform workers about them, and to ensure appropriate protection against risks arising directly from the provisions of the labour codes of each Member State.

## 10.3 RULES FOR PREPARING A MICROBIOLOGICAL EXPERT REPORT ON THE BUILDING

Microbiological corrosion always leads to significant and costly damage, to decrease and/or loss in functional conditions of the building and, in extreme cases, even to its complete decommissioning. Therefore, the easiest way to counteract the microbiological corrosion is to properly design and maintain the buildings in accordance with the laws of building physics, i.e. limiting the possibility of growth of harmful microorganisms by eliminating physical phenomena facilitating bio-corrosion, such as water capillary action, permeation or condensation within the building envelope. However, if dampness and colonisation of construction materials by microorganisms occurs, these processes can practically only be stopped by immediate intervention. Only with professional diagnosis and appropriately selected technology is it possible to remove the causes and symptoms of bio-corrosion in a building [Klemm 2010].

The commencement of remediation work on a biologically corroded facility should be preceded by an appropriate microbiological expert report. Its content must be adapted to the legal and economic conditions in the respective country, including standards, regulations and guidelines, and must take into account the economic conditions of building renovation and the availability of disinfection methods. The analysis can be performed by specialists with knowledge of microbiological corrosion of construction materials, who are authorised to perform independent technical functions in building construction [Ważny and Karyś 2001]. In many countries (e.g. in the US), companies that provide such expertise and perform construction work in biologically corroded buildings must have an appropriate licence [Major and Boese 2017]. It should be emphasised that a prerequisite for the effectiveness of remediation works is the proper identification of the cause of biological corrosion of the evaluated facility. Therefore, in addition to the assessment of the damage caused by microorganisms, the expert report should include the identification of any defects and anomalies in the construction and finishing elements of the building, which may lead to the formation and development of biological corrosion. Such a document should include the elements listed below [Ważny and Karyś 2001].

### 10.3.1 Initial Part of the Expert's Work

This part should describe the purpose of a microbiological expert report. It should provide an overview of available documents, including project documentation, cost estimates and previous expert reports (if any, including reasons for them being carried out). The initial part should also include a general description of the building, including information about the number of storeys, basement conditions, roof and wall structure, finishing elements, types of materials used for installations etc.

An important element of the report is to carry out an on-site inspection. A detailed description of the technical conditions indoors (including defects and anomalies in construction and finishing elements) and outdoors in relation to building structure (with a description of the condition of the facade, water drainage facilities, insulation etc.) must be made. An on-site inspection is the most important step in identifying the problem of dampness and bio-corrosion in a building. It enables the assessment of the extent of damage caused by water and microorganisms growing on construction and finishing materials. During an on-site inspection, special attention should be paid to typical symptoms of bio-corrosion, i.e. increased humidity of the air and construction materials; unpleasant mouldy odours; stains on walls; salt efflorescence and fungal coatings (white coatings, dark spots), paint peeling etc. Environmental sampling in the facilities is usually not necessary; however, it can be helpful in some cases, e.g. when users report health problems or when bio-corrosion affects sites of particular importance (e.g. monuments). It is essential to measure the temperature and humidity of the air and materials (e.g. walls) every time. Humidity measurements of materials are usually performed using dielectric and electrofusion, gravimetric or isotopic methods [Ważny and Karyś 2001].

Decisions on appropriate remediation can generally be taken on the basis of a thorough on-site inspection. This part of the expert's work also includes precise photographic documentation presenting the actual state of the facility with its hazards, marking the places where excavations (e.g. ceilings, foundations) and sampling (scrapers, drilling samples, swabs etc.) were made. It is also necessary to attach drawings or descriptions of the excavations. An important element is also to describe the characteristics of the construction materials in detail and to conduct energy audit calculations. This part also includes information from interviews with users of the building, its designers or contractors.

### 10.3.2 Identification of the Problem

On the basis of the on-site inspection and the collected data, the reasons for the increased level of humidity and bio-corrosion of the facility under investigation should be determined. The proper identification of the source of dampness and contamination is crucial for the application of appropriate remediation methods. If microbiological analyses have been carried out, the possible effects of the identified microorganisms on human health and on the technical condition of construction and finishing materials must be stated in this part of the expert opinion.

### 10.3.3 Recommendations

This section should contain a description of methods suitable to remove the causes of identified hazards, as recommended by the expert. Plans and drawings should indicate where the contamination occurs and where the remediation work should be performed.

## REFERENCES

Brennan, T., G. Cole, and B. Stephens. 2015. Report to EPA on guidance documents to safely clean, decontaminate, and reoccupy flood-damaged houses. https://www.epa.gov/indoor-air-quality-iaq/report-flood-related-cleaning. [accessed October 11, 2019].

Directive 2000/54/EC of the European Parliament and of the Council of 18 September 2000 on the Protection of Workers from Risks Related to Exposure to Biological Agents at Work. *Official Journal of European Communities L.* 262/21, Brussels (with subsequent amendments: Commission Directive (EU) 2019/1833 of 24 October 2019 amending Annexes I, III, V and VI to Directive 2000/54/EC of the European Parliament and of the Council as Regards Purely Technical Adjustments. *Official Journal of European Communities L.* 279/54).

Directive 2010/31/EU of the European Parliament and of the Council of 19 May 2010 on the energy performance of buildings. *Official Journal of European Communities L* 153, 18 June 2010, p. 13–35.

DOHMH [New York City Department of Health & Mental Hygiene]. 2008. *Guidelines on Assessment and Remediation of Fungi in Indoor Environments.* New York, NY: New York City Department of Health & Mental Hygiene. https://www1.nyc.gov/assets/doh/downloads/pdf/epi/epi-mold-guidelines.pdf. [accessed October 11, 2019].

EN 335:2013-07. 2013. Durability of wood and wood-based materials: Classes of use: Definitions, application to solid wood and wood-based materials.

EN 350:2016-10. 2016. Durability of wood and wood-based materials: Testing and classification of the durability of wood and wood-based materials against biological agents.

EPA [United States Environmental Protection Agency]. 2008. Mold remediation in schools and commercial buildings. https://www.epa.gov/sites/production/files/2014-08/documents/moldremediation.pdf. [accessed October 11, 2019].

EPA, HUD, NIH and OSHA [United States Environmental Protection Agency, Department of Housing and Urban Development, National Institutes of Health and Occupational and Safety and Health Administration]. 2015. Worker and employer guide to hazards and recommended controls. https://www.epa.gov/mold/worker-and-employer-guide-hazards-and-recommended-controls. [accessed October 11, 2019].

IOM [Institute of Medicine]. 2004. *Damp Indoor Spaces and Health.* Washington, DC: The National Academies Press.

Journal of Laws of 1994, no. 89, item 414. Ustawa z dnia 7 lipca 1994 r. Construction law.

Klemm, P. 2010. *Budownictwo ogólne, T. 2: Fizyka budowli*. Warsaw: Arkady.

Major, J. L., and G. W. Boese. 2017. Cross section of legislative approaches to reducing indoor dampness and mold. *J Public Health Manag Pract* 23(4):388–395.

MEDEP [Maine Department of Environmental Protection]. 2007. Report of the mold in Maine buildings task force. https://irp-cdn.multiscreensite.com/c4e267ab/files/uploaded/JGp5PpWRHuxAAVoxmSWg_Maine_Report_of_the_Mold_in_Maine_Buildings_Task_Force_2007.pdf. [accessed October 11, 2019].

NCHH [National Center for Healthy Housing]. 2008. Creating a healthy home: A field guide for clean-up of flooded homes. https://nchh.org/resource/creating-a-healthy-home-a-field-guide-for-clean-up-of-floodefd-homes. [accessed October 11, 2019].

New York State Labor Law. 2015. Licensing of mold inspection, assessment and remediation specialists and minimum work standards. https://www.nysenate.gov/legislation/laws/LAB/A32. [accessed October 11, 2019].

OSHA [Occupational Safety and Health Administration]. 2003. A brief guide to mold in the workplace. *SHIB 03.10.10*. http://www.osha.gov/dts/shib/shib101003.html. [accessed October 11, 2019].

PN EN 206+A1:2016–12. 2016. Beton: Wymagania, właściwości, produkcja i zgodność.

Regulation (EU) 305/2011 of the European Parliament and of the Council of 9 March 2011. Laying Down Harmonised Conditions for the Marketing of Construction Products and Repealing Council Directive 89/106/EEC. *Official Journal of European Communities* L 88, 4 April 2011, p. 5–43.

Rozporządzenie Ministra Transportu, Budownictwa i Gospodarki Morskiej z dnia 5 lipca 2013 r. zmieniające rozporządzenie w sprawie warunków technicznych, jakim powinny odpowiadać budynki i ich usytuowanie. Dz. U. z 2013 r., poz. 926.

Urban Green Building Council. 2010. NYC Green Codes Task Force: A report to Mayor Michael R. Bloomberg & Speaker Christine C. Quinn. http://urbangreencouncil.org/sites/default/files. [accessed October 11, 2019].

Ważny, J., and J. Karyś. 2001. *Ochrona budynków przed korozja biologiczną*. Warsaw: Arkady.

WHO [World Health Organization]. 2009. *Guidelines for Indoor Air Quality: Dampness and Mould*. Copenhagen: WHO Regional Office for Europe.

# 11 Microbiological Corrosion of Buildings in Everyday Practice – Examples

*Małgorzata Gołofit-Szymczak*

**CONTENTS**



## 11.1 RESIDENTIAL BUILDINGS

A significant number of microorganisms are found in residential buildings, in different rooms. The levels of bacterial concentrations in the air of apartments range from 140 CFU/m$^3$ to 1200 CFU/m$^3$. In the case of old houses, the concentration of bacteria can even rise to over 2000 CFU/m$^3$. Values of fungal concentrations can range from 56 CFU/m$^3$ to 800 CFU/m$^3$. In water-damaged buildings (after floods, hurricanes, water and sewage system failures), the fungal airborne concentrations may even reach values exceeding 5000 CFU/m$^3$ [Kowalski 2006; Rao et al. 2007; Riggs et al. 2008; Barbeau et al. 2010].

From the surface of household appliances, in premises such as kitchens and bathrooms, *Bacillus* and coagulase-negative bacteria of *Micrococcaceae* genera are isolated in the greatest number. In the air of these rooms, however, microorganisms from *Micrococcus*, *Staphylococcus*, *Aerococcus*, *Moraxella*, *Pseudomonas* genera and actinomycetes are the most prevalent. Appliances such as wash-basins, sinks and kitchen drainers are also emitters of bacterial aerosol, mainly from the *Enterobacteriaceae* family [Nevalainen 1989].

Room carpets, by accumulating dust particles together with microbiological particles, become an abundant source of microbial emissions over time. Moulds, mesophilic and thermophilic bacteria, endotoxins and (1→3)-β-D-glucans are the most common microbial components present in them [Cole et al. 1993]. Studies have shown that the concentrations of fungi and bacteria above the surface covered with carpet are significantly higher than those above the floor without carpets [Pope et al. 1993].

The use of wood, cardboard or cork as finishing or insulating materials may cause additional microbial and bacterial toxin emissions in an indoor environment [Cox and Wathes 1995; Pope et al. 1993]. The analysis of samples of these materials shows that under favourable humidity conditions, they may contain up to $10^7$ CFU/g of microorganisms, and endotoxin concentration may reach $10^5$–$10^6$ EU/g [Dutkiewicz 1989]. The microorganisms colonising these materials include Gram-negative bacteria (*Pantoea agglomerans*, *Agrobacterium radiobacter*, *Pseudomonas fluorescens*, *Xanthomonas maltophilia*, *Acinetobacter calcoaceticus*), corynebacteria, *Bacilli*, yeasts (e.g. of *Candida*, *Cryptococcus*, *Rhodotorula* genera) and moulds (e.g. *Penicillium* spp., *Aureobasidium pullulans*, *Aspergillus fumigatus*, *Trichoderma* spp.) from wood [Dutkiewicz et al. 1992], as well as *Aureobasidium* spp. and *Streptomyces* spp. from cork [Pope et al. 1993].

Precast concrete, which is used on a large scale in residential buildings, due to freezing and dampness can also become an emitter of bioaerosol pollutions. As a result of these processes, the 'moulding of buildings', i.e. the mass growth of moulds (mainly of *Ascomycetes* and *Phylomycetes* genera, less frequently *Deuteromycetes*) on the surface and inside the building partitions causes spore contamination of the interiors, both of the air and walls and equipment surfaces [Doleżal 1993].

Ventilation and humidifying systems are very often the source of bioaerosols. Standing water used in the operation of these devices becomes an excellent habitat for the development of microorganisms over time [Lewis et al. 1990; Nevalainen 1989; Pope et al. 1993]. *Pseudomonas aeruginosa*, *Flavobacterium* spp., *Acinetobacter* spp., *Alcaligenes* spp. and thermophilic actinomycetes are potentially pathogenic organisms that are associated with the presence of ventilation ducts and air humidifiers [Hung et al. 2005; Flannigan et al. 1991; Morey and Feeley 1990].

Soil, which is often used in home pot plants, can also cause microbial contamination of indoor air. Among these pollutants are usually chemoorganotrophic bacteria and fungi. One gram of biologically active soil may contain, apart from other non-biological components, millions of bacterial and fungal cells [Petrycka 1993].

Household waste is a significant reservoir of microorganisms in the indoor environment. Any waste containing organic substances is an ideal environment for the growth of many microorganisms, and its number in household waste is significant, reaching $10^8$–$10^9$ CFU/g. The qualitative composition of the microbiota depends on the type of waste collected. And so, in segregated biological waste, the most prevalent are bacteria from *Enterococcus*, *Escherichia* and *Pseudomonas* genera; in unsorted household waste, the most numerous are bacteria from *Salmonella*, *Escherichia*, *Staphylococcus*, *Yersinia*, *Streptococcus* and *Aspergillus* genera; in mixed waste containing recycled paper, predominant are microorganisms from *Staphylococcus*, *Streptococcus*, *Acinetobacter*, *Enterobacter*, *Citrobacter*, *Hafnia*, *Klebsiella*, *Proteus*, *Salmonella*, *Serratia*, *Aeromonas*, *Pseudomonas* and *Kluyvera* genera [Jager and Eckrich 1997].

## 11.2 OFFICE BUILDINGS

For over a dozen years the dynamic development of office construction has been observed all over the world. This is followed by an increase in the number of people employed in premises adapted for office work. These workers are one of the more numerous occupational groups in all sectors of public and private activity, representing about 30% of the total number of employees. Epidemiological data indicate that every third office worker suffers from health problems which are caused by poor quality air supplied to this type of interior. Office workers often complain about fatigue, dyspnoea, headaches and dizziness, irritability, reduced ability to concentrate, memory disorders, irritation of the conjunctiva and upper respiratory tract and skin lesions. These syndromes of unspecific, subjective symptoms arising from indoor exposure have been described as sick building syndrome (SBS), tight building syndrome (TBS) or diseases occurring as a result of being in an excessively polluted indoor environment called building related illnesses (BRI).

The air in offices can be contaminated by both microorganisms (viruses, bacteria, fungi) and the substances they produce and secrete (e.g. glucans, mycotoxins, MVOCs). In the air of the office premises, the most frequently identified viruses include orthomixoviruses (e.g. influenza viruses), picornaviruses (most commonly rhinoviruses), coronaviruses, adenoviruses, paramixoviruses (e.g. parainfluenza viruses) and caliciviruses (including noroviruses) [La Rosa et al. 2013]. In an American study by Prussin et al. (2015), conducted in office premises, the average concentration of viruses in the air was $4.9 \times 10^5$ particles/m$^3$.

The air of office premises may also contain several dozen species of bacteria, which usually account for 60–90% of the total number of air-polluting microorganisms. The majority of them are Gram-positive cocci from *Micrococcus*, *Staphylococcus* and *Streptococcus* genera, Gram-positive rods from *Bacillus* genus and Gram-negative bacteria (mainly from *Enterobacter*, *Pseudomonas*, *Serratia*, *Klebsiella*, *Proteus* genera). Water supply networks, warm water systems, humidifiers, sprinklers, air-conditioning and refrigerating equipment are places where *Legionella* bacteria find good conditions for growth. The infection is caused by inhalation of contaminated water aerosols with a droplet diameter of 0.2–5 μm [Carducci et al. 2010].

Fungi are a particular problem among microbiological factors in the air of office buildings. The toxic and allergenic compounds they secrete can cause a number of adverse health effects in humans. Studies carried out in Europe over the last dozen or so years have shown that about 400 species of fungi may be present in the indoor air. Of these, the most frequently and numerously represented are mould species from *Alternaria*, *Cladosporium* (including *C. sphaerospermum*, *C. cladosporioides*, *C. herbarum*), *Penicillium* (*P. chrysogenum*, *P. viridicatum*, *P. expansum*), *Aspergillus* (including *A. niger*, *A. flavus*), *Rhizopus nigricans*, and *Mucor* genera and yeasts, mainly species from the *Candida* genus [Gots et al. 2003; Herbarth et al. 2003;

Crook and Burton 2010]. The occurrence and growth of some moulds is associated with the release of allergens, mycotoxins, volatile organic compounds and glucans into the environment [Rylander and Peterson 1994; Midtgaard and Poulsen 1997]. Species of *Alternaria*, *Cladosporium*, *Aspergillus*, *Penicillium*, *Trichoderma* and *Mucor* genera are the most important cause of mould allergy.

Quantitative analysis of bioaerosol showed that the concentrations of bacterial and fungal aerosols, measured by the volumetric method in office spaces, are usually in the range between $10^1$–$10^3$ CFU/m$^3$ [e.g. Brickus et al. 1998; Bonetta et al. 2010; Gołofit-Szymczak and Górny 2010; Tseng et al. 2011].

Most office spaces are equipped with either mechanical ventilation or air-conditioning systems. This type of system installed in a building and working efficiently, makes the free migration of bioaerosol from the outdoor to indoor environment virtually impossible. The air entering the system as well as the air distributed to individual premises, is treated (usually through its filtration) in order to eliminate the pollutants it contains. An improperly maintained ventilation system can be a source of additional contamination of the air flowing through the ventilation ducts as a result of secondary particle dusting [Sundell et al. 2011].

## 11.3 INDUSTRIAL FACILITIES

### 11.3.1 Waste and Wastewater Management

One of the activities with particularly high risks from exposure to harmful biological agents is waste and wastewater management. Biological agents may occur in waste and wastewater itself and be released into the air, water and soil. Municipal waste, due to the fact that the organic phase has a large share in its structure, provides an excellent culture medium for microorganisms. There are high concentrations of microorganisms, carried by organic dust, in landfills, sorting plants, incineration and composting plants. The average concentrations of organic dust during manual waste sorting are between 0.08–33.4 mg/m$^3$. The concentrations of fungal aerosol vary between $10^4$–$10^5$ CFU/m$^3$, whereas those of bacterial aerosol range from $10^2$ to even $10^9$ CFU/m$^3$. The highest concentrations of bacteria in the air were recorded in waste composting plants [Kozajda et al. 2009]. More than 100 bacterial species have been identified in these types of industrial facilities, mainly from *Acinetobacter*, *Bacillus*, *Clostridium*, *Enterobacter*, *Escherichia*, *Micrococcus*, *Listeria*, *Pseudomonas*, *Serratia*, *Staphylococcus*, *Streptomyces* and *Thermoactinomyces* genera. The fungi are dominated by species from *Aspergillus*, *Penicillium* and *Candida* genera. Special attention should be paid to Gram-negative bacteria and endotoxins they produce. They are detected at all stages of municipal waste management. Their range varies, depending on the activities and weather conditions, from 0.01 ng/m$^3$ to 600 ng/m$^3$ [Szadkowska-Stańczyk 2007].

Dust and drop aerosol, formed during numerous technological processes in mechanical, biological and sludge treatment, can contain microorganisms in a very wide range of concentrations. During the wastewater aeration process, the aerosol over sewage sludge pools may contain microorganisms in the concentration range between $10^1$–$10^4$ CFU/m$^3$ [Kowalski 2006]. The results of Laitinen's study

[1999] indicate that the highest concentrations of bioaerosol ($2.5 \times 10^4$ CFU/m$^3$) were found in the part of the treatment plant where sludge was processed, while the lowest concentrations ($5.6 \times 10^2$ CFU/m$^3$) were found in the sewage pumping station. The most frequently determined bacterial species included *Klebsiella pneumoniae*, *Escherichia coli*, *Enterobacter agglomerans*, *Aeromonas hydrophila*, various species of *Pseudomonas*, *Micrococcus*, *Streptococcus*, *Bacillus*, *Proteus* and *Staphylococcus* genera. Among the most common fungal species were those from the *Aspergillus*, *Cryptococcus* and *Candida* genera [Cyprowski and Krajewski 2003]. The presence of Gram-negative bacteria in sewage treatment plants is always associated with the presence of endotoxins. The highest values of endotoxin concentration, reaching 5931 ng/m$^3$ were recorded during the processing of sewage sludge [Darragh et al. 1997].

### 11.3.2 Power Industry

The main hazard to workers in this sector is organic dust, which is generated in the processing and use of biomass. Exposure to harmful biological agents depends both on the type of raw material used and on transport and storage conditions. Fresh biomass can contain up to 2,000 species of bacteria and fungi and the content of microorganisms, e.g. in wood chips, can reach $10^5$ CFU/g. Among the factors of microbial origin with immunotoxic properties found in the biomass dust are bacterial endotoxin produced by Gram-negative rods of *Enterobacteriaceae* family (e.g. *Rahnella* spp., *Pantoea* spp., *Enterobacter* spp., *Proteus* spp.), *Pseudomonas* and *Alcaligenes* genera as well as thermophilic (*Saccharomonospora viridis*, *Thermoactinomyces vulgaris*) and mesophilic (*Streptomyces* spp.) actinomycetes. Moulds of *Aspergillus* (*A. fumigatus*), *Penicillium*, *Alternaria*, *Scopulariopsis* and *Trichoderma* genera are also a serious hazard in biomass processing. The analysis of bioaerosol concentrations at workstations in facilities processing biomass for energy purposes showed that concentrations of bacteria, fungi and endotoxins in the air may reach $10^7$ CFU/m$^3$, $10^6$ CFU/m$^3$ and $10^5$ JE/m$^3$, respectively [Ławniczek-Wałczyk et al. 2012].

### 11.3.3 Machine Industry

Exposure to harmful biological agents in the machine industry sector is linked to the presence of endotoxins and bacteria allergens. Gram-negative bacteria, which grow abundantly in used oils and oil–water emulsions used for cooling, lubrication and cleaning of machinery and equipment could be a component of the so-called 'oil mist' and occur near machines at a concentration of $10^3$–$10^8$ CFU/m$^3$ [Cox and Wathes 1995]. These bacteria are predominantly of the *Pseudomonadaceae* family and *Pseudomonas fluorescens* has been identified as the probable cause of a new subunit of machine operator's lung [Bernstein et al. 1995]. Nontuberculous mycobacteria (*Mycobacterium immunogenum*) may occur in oils and 'oil mist'. They are the cause of hypersensitivity pneumonitis (HP) in employees of the machine industry [Beckett et al. 2005, Tillie-Leblond et al. 2011]. Another species of mycobacteria from the *Mycobacterium avium* complex group may be the cause of a specific form of HP known as 'hot-tub lung', which occurs in people inhaling the aerosol created by spraying hot water [Beckett et al. 2005]. The concentrations of microorganisms in the halls of the machine industry are dependent on the processes (grinding, cutting), during which oil mist is released as a result of the rotation of machine parts. The concentration of bacteria can then vary between $10^1$–$10^5$ CFU/m$^3$ [Cyprowski et al. 2016].

### 11.3.4 Woodworking Industry (Joineries, Sawmills, Pellet Factories, Pulp and Paper Mills)

A significant number of microorganisms are found in rooms where woodworking is performed. Average concentrations of bacteria in Polish joineries and sawmills were 737 CFU/m$^3$ and 953 CFU/m$^3$, respectively. The average concentrations of fungi were 135 CFU/m$^3$ in joineries and 211 CFU/m$^3$ in sawmills. The dominant microbiota consists here of Gram-positive cocci, Gram-positive endospore-forming and nonsporing bacilli and Gram-positive rods. In sawmills, on the other hand, moulds (mainly from *Aspergillus* and *Cladosporium* genera) and nonsporing Gram-positive rods are the most prevalent. In recent studies in wood pellet plants, the average concentration of bacteria in the air was 2018 CFU/m$^3$, while that of fungi was 564 CFU/m$^3$ [Górny et al. 2019]. In earlier Polish studies [Prażmo et al. 2000; Dutkiewicz et al. 2001a, 2001b], the concentration ranges of Gram-negative bacteria in sawmills, fibreboard and chipboard factories were between $10^1$–$10^4$ CFU/m$^3$ (with the predominance of *Rahnella*, *Enterobacter* and *Pantoea* genera), $10^3$–$10^4$ CFU/m$^3$ for the total bacteria (with predominance of coryneform and Gram-negative bacteria) and $10^1$–$10^4$ CFU/m$^3$ for fungi (with the predominance of *Aspergillus* and *Penicillium* genera). The comparison of these results with the above-listed studies by Górny et al. [2019] shows a significant improvement in occupational hygiene conditions in the wood industry over the last 20 years. Nevertheless, at some workplaces in this industry, significant concentrations of allergenic microorganisms may occur, as recently evidenced by Mackiewicz et al. [2019] on the cases of HP among furniture factory workers as a result of exposure to *Pantoea agglomerans* and *Microbacterium barkeri*.

The main source of microbiological contamination in pulp and paper mills is considered to be water, raw materials used, machine parts and indoor air. Paper machines provide an ideal environment for the development of microorganisms that can enter the system along with water, raw materials and additives. As a plant raw material, paper can be colonised and subject to biocorrosion at almost every stage of its production and processing. Therefore, microbiological contamination may occur in the paper production facilities. Some species of microorganisms growing on finished paper products can affect their quality. Microorganisms occurring during paper processing include species from *Bacillus*, *Bacteroides*, *Cellulomonas*, *Cellvibrio*, *Clostridium*, *Cytophaga*, *Desulfovibrio*, *Enterobacter*, *Escherichia*, *Eubacterium*, *Klebsiella*, *Microbispora*, *Micromonospora*, *Pantoea*, *Propionibacterium*, *Pseudomonas*, *Rahnella*, *Sporocytophaga*, *Streptomyces*, *Alternaria*, *Aspergillus*, *Candida*, *Chaetomium*, *Cladosporium*, *Epicoccum*, *Fusarium*, *Mucor*, *Myrothecium*, *Penicillium*, *Rhodotorula*, *Scopulariopsis*, *Stachybotrys*, *Trichoderma* and *Verticillium* genera [Flemming et al. 2013; Desjardins and Beaulieu 2003; Prażmo et al. 2003].

## 11.4 AGRICULTURAL FACILITIES

The most important sources of harmful biological agents in agriculture include farm animals as well as animal housing, arable crops and soil. Regardless of whether the agricultural production process takes place on large or small farms, the nature of the work is related to the long-term presence of a highly biologically polluted environment. There may be numerous microorganisms responsible for zoonoses in secretions, excretions and animal tissues. Arable plants (cereals, herbs) are an abundant source of microorganisms that can develop on growing plants (e.g. *Pantoea agglomerans*) or on stored products (e.g. moulds of *Aspergillus*, *Penicillium*, *Fusarium* or *Stachybotrys* genera and actinomycetes) [Dutkiewicz 1997].

Despite the above, the main hazard in the agricultural sector is organic dust. It contains feed and faeces particles, feather, desquamated skin, bedding etc. from the animal housing facilities, and the microorganisms (bacteria, fungi and viruses) that develop on them. The works during which the largest amounts of organic dust are released include field crop harvesting, grain threshing, flax breaking, herb cleaning and cattle, pig and poultry breeding [Zuskin et al. 1995]. Farmers and workers in the agri-food industry are exposed to the inhalation of organic dusts that contain large quantities of microorganisms and compounds with immunotoxic properties (endotoxins, glucans) produced by them, as well as plant and animal allergens.

The level of microbiological contamination of the air and surfaces in animal and plant production facilities is variable and depends on a number of factors, the most important of which are: the type of activity (plant, animal), species and age of livestock, their number and concentration, frequency of cleaning and disinfection of livestock facilities. Animal buildings may also be a significant source of bioaerosol [Seedorf 2004]. Particularly noteworthy are those livestock buildings where the animals, their secretions and excretions and the specific microclimate created by them can be an excellent place for the settlement and proliferation of various microorganisms [Popescu et al. 2014]. The presence of a significant number of microorganisms in such interiors may threaten the health of both humans and farm animals [Szadkowska-Stańczyk et al. 2010; Mc Donnell et al. 2008]. The risk of contamination of the air, walls, flooring and bedding in livestock areas increases when the breeding environment creates a humid and warm microclimate. *Enterobacteriaceae*, *Pseudomonas* and *Acinetobacter* multiply particularly easily under these conditions. The condition of bedding has also a significant impact on air quality in livestock buildings. Busato et al. [1999] performed microbiological evaluation of bedding and isolated species from *Salmonella*, *Campylobacter*, *Escherichia*, *Yersinia*, *Staphylococcus*, *Streptococcus* and *Pseudomonas* genera.

Significantly higher concentrations of bacterial aerosol are observed in facilities where animals are kept (chicken coops, piggeries) than in those where plant production is carried out (barns, granaries) [Seedorf 2004]. The highest bacterial concentrations were observed in the chicken buildings. It was found that during an 8-week rearing period of broiler chickens (with the stocking density of 15 chickens per 1 $m^2$), regardless of efficient ventilation, about 1 kg of dust/$m^2$ is deposited [Ławniczek-Wałczyk et al. 2013]. The concentration of microorganisms in the livestock housing and during pre-treatment of plant raw materials (e.g. grain, herbs, flax) is usually $10^5$–$10^6$ CFU/$m^3$ [Dutkiewicz 1989].

In the air of the pig house, moulds of *Absidia*, *Alternaria*, *Cladosporium*, *Rhizopus*, *Scopulariopsis*, *Penicillium*, *Aspergillus* and *Mucor* genera and numerous yeasts of the *Candida*, *Cryptococcus*, *Saccharomyces* genera may be present [Martin et al. 1996]. According to Masclaux et al. [2013], pathogenic staphylococci are also often isolated from the air in pig buildings.

Preventive measures taken to eliminate or reduce microorganisms in agricultural production premises should include the application of appropriate procedures for cleaning and disinfection of breeding facilities, the monitoring of zoonoses, periodic deratisation and disinsectisation treatments, reduction of pollination, e.g. by improving the ventilation system, and safe and efficient waste disposal and treatment.

## 11.5 HEALTHCARE FACILITIES

Healthcare facilities are a specific environment in which patients, workers and visitors are exposed to harmful biological agents. The sources of microorganisms can be personnel, equipment subjected to an improper disinfection process, patients' endogenous microbiota and carriers of infectious diseases, as well as air and water. Healthcare workers (physicians, nurses, medical technician, orderlies as well as workers in diagnostic and scientific laboratories) are exposed to infectious agents of known and sometimes unknown biological origin. The highest risk level occurs among the staff of infectious, surgical, pulmonological, haematological, gynaecological-obstetrical, intensive care and haemodialysis wards, as well as among dentists, paediatricians, family doctors and ambulance staff, blood donation stations and laboratory workers. Hepatitis B, C, D and G viruses, transmitted mostly by blood, as well as respiratory diseases and tuberculosis mycobacteria are particularly dangerous. Workers in animal quarters (vivaria) at scientific institutes and pharmaceutical plants are exposed to strong allergens present in rodent excretions [Patterson et al. 1985].

In recent years, there has been a growing worldwide interest in sanitary and epidemiological services in the problem of threat to human health and life caused by bacteria of the *Legionella* genus. Hospitals are among the places where the probability of an increase in bacterial counts in water and air-conditioning systems and the possible risk for people staying there and particularly susceptible to infection is very high. Sources of *Legionella* infection in healthcare facilities may include warm water systems, air-conditioning systems, but also air humidifiers, swimming pools, shower spray plates, cooling towers, washrooms, dental turbines, respiratory support devices, dialysers etc. Improperly maintained (no systematic cleaning or disinfection) ventilation systems can be a source of contamination of air flowing through ventilation ducts due to secondary dusting [Turetgen and Cotuk, 2007]. In a study conducted by Farhat et al. [2018], all samples taken in the hospital air-conditioning system (water from the cooling tower) confirmed *Legionella pneumophila* rods presence.

## 11.6 HISTORIC FACILITIES: MUSEUMS AND ART CONSERVATION LABORATORIES

Collection storage facilities are a specific microenvironment and its nature depends on many factors. It consists, among others, of the type of premises and their technical condition, the nature of the collections gathered and the manner and conditions of their storage. Collections are usually stored and conserved in rooms adapted into storage and art conservation laboratories, located away from exhibition halls, often in basements or at the attics. The technical condition of these rooms favours the development of microbiota [Abe 2010; Ascione et al. 2009; Sterflinger 2010]. Materials used in creative processes resulting in works of art or cultural monuments are usually organic and of natural animal and plant origin (wood, paper, leather and plant fibres, and often containing animal or plant glue). Despite many of their advantages used in the creation process, they have limited durability and are susceptible to biodeterioration caused by bacteria and fungi [Sterflinger 2010]. The variety of collections stored in common warehouses does not allow the creation of climate comfort for all groups of materials used to make the collected objects. Optimal values of relative air humidity and temperature in the storage process of some materials can be dangerous for others. Moreover, as a result of various unanticipated events, traces of biological corrosion may appear on the surface of objects. These objects, if not disinfected, are then a potential source of infection for other items in the collection. Cultural institutions rarely have the opportunity to carry out full and systematic disinfection of collections, and if this is done, it is usually performed for individual objects that will return to the contaminated environment anyway [Mandrioli et al. 2003].

Microorganisms can migrate to museum premises and art conservation laboratories together with workers or visitors on their bodies or clothes, brought in items or with the outdoor air through 'natural gates' (doors and windows). Moreover, a polluted or poorly functioning ventilation or air-conditioning system causing dynamic changes in the air temperature and humidity may promote microbiological contamination of such interiors and, by that, pollution of art objects or archives [Aira et al. 2007; Borrego et al. 2012; Joblin et al. 2010].

Biodeterioration is a well-known phenomenon that poses the greatest threat to various types of cultural monuments. Microorganisms, which are constantly present in both outdoor and indoor environments, pose the risk of damage and even total destruction of all types of art works, from sculptures, through pictures, paintings and photographs, to literary monuments. This process is particularly evident when the environmental conditions in which the artworks are kept or stored are conducive to the development of harmful microbial agents [Fazio et al. 2010; Mandrioli et al. 2003]. Due to almost free access to nutrients contained in the material of which these objects are made and the availability of moisture, microorganisms are able, albeit slowly, to destroy them systematically. Destructive changes are usually caused by a wide spectrum of chemical compounds produced by microorganisms, including enzymes, organic and inorganic acids, vitamins, amino acids, some purines and other organic compounds such as methane, dimethyl sulphide, antibiotics, toxins, chelating compounds or pigments [Mandrioli et al. 2003; Strzelczyk and Karbowska-Berent 2004].

Some bacteria (e.g. *Streptomycetes*) and numerous moulds (e.g. *Botrytis, Trichoderma, Chaetomium, Alternaria, Ulocladium, Aspergillus, Penicillium*) have strong cellulolytic properties and can significantly damage historically important objects such as books and other paper documents, textiles, furniture and furnishings, paintings and sculptures. Biodegradation of protein-derived materials such as leather or parchment can be performed by both bacterial (*Bacillus*, *Pseudomonas*, *Clostridium*, *Streptomycetes*) and fungal (*Mucor*, *Chaetomium*, *Aureobasidium*, *Gymnoascus*, *Trichoderma*, *Verticillium*, *Epicoccum*) species [Abdulla et al. 2008; Mesquita et al. 2009; Valentin 2003].

Fresco paints and pigments, especially those using casein, egg and emulsion tempers, methylcellulose, polyvinyl alcohol and polyvinyl acetate, are also susceptible to biological degradation processes. For example, Guglielminetti et al. [1994], when studying the 16th-century frescoes of the monastery in Assisi, isolated 21 fungal genera (including *Cladosporium*, *Penicillium*, *Alternaria*, *Chaetomium* and *Acremonium*), the presence of which caused the frescoes' biodeterioration.

## 11.7 SPECIAL FACILITIES: LIBRARIES AND ARCHIVES

Due to their strictly defined function, the rooms in which book collections are stored represent a specific working environment. The hygiene condition of these premises is often poor. They are usually located on the lowest (basement) or highest (attic) floors, the collections stored in them are usually rarely cleaned and the ventilation or air-conditioning systems, if any, are often inadequately maintained. Structural defects of buildings as well as failures of water and sewage systems make both the storage rooms and the book collections or archival materials stored therein vulnerable to water damage. Over time, the microbiologically contaminated collections become a source of emission of harmful agents and therefore pose a serious threat to the health of people staying or working in such polluted environment [Hempel et al. 2014; Pinheiro et al. 2011; Haleem Khan and Mohan Karuppayil 2012].

The high content of cellulose as well as protein and other organic and inorganic substances found in bookbinding mortars, library and archive collections (paper, papyrus, parchment, cardboard, photographs) is an ideal culture medium for various microorganisms [Borrego et al. 2012; El-Nagerabi et al. 2014]. The accumulation of a large number of items on shelves and bookstands can also limit the flow of air, which further increases the possibility of free development of microbiota on the surface of stored materials [El-Nagerabi et al. 2014].

Most of the microorganisms colonising archive and library collections are saprophytes, which means that they extract nutrients from dead moist matter such as wood, paper, paints, glues, dust etc. They can also grow with equal success on surfaces composed of moist inorganic matter. Among the fungi contaminating the stored objects are species from *Aspergillus*, *Penicillium*, *Trichoderma*, *Botrytis*, *Alternaria*, *Stemphylium*, *Mucor*, *Chaetomium*, *Aureobasidium*, *Gymnoascus*, *Verticillium* and *Epicoccum* genera. Bacteria of *Streptomycetes*, *Flavobacterium*, *Bacillus*, *Pseudomonas* and *Cellulomonas* genera are also often isolated from the surface of the collected items [El-Nagerabi et al. 2014; Maggi et al. 2000; Zielińska-Jankiewicz et al. 2008]. The concentrations of microorganisms in the air of libraries and archives may vary between 3.2–7.2 × $10^2$ CFU/$m^3$ [Skóra et al. 2015].

Numerous microbiological catastrophes have occurred in libraries around the world during the 20th century. One example of such a disaster took place in the Arsenal Library in Paris. After a serious failure of the air-conditioning system, 240,000 volumes were contaminated. The presence of moulds (*Aspergillus wentii*,

*Chaetomium crispatum*, *Fusarium* spp. and *Penicillium* spp.) was detected on both the book covers and bookshelves. Disinfection of this book collection with ethylene oxide took 16 months [Coron and Lefevre 1993]. Apart from such drastic action, the primary factor in protecting library and archive collections from upper biodegradation is internal air control. According to Zyska [1997], the optimal air temperature for books stored in that way is 15°C and the best relative humidity of the air is 45%.

## REFERENCES


Abdulla, H., E. May, M. Baghat, and A. Dewedar. 2008. Characterization of Actinomycetes isolated from ancient stone and their potential for deterioration. *Pol J Microbio* 57:213–220.

Abe, K. 2010. Assessment of the environmental conditions in a museum storehouse by use of a fungal index. *Int Biodeterior Biodegrad* 64(1):32–40.

Aira, M. J., V. Jato, A. M. Stchige, R. J. Rodriguez-Rajo, and E. Piontelli. 2007. Aeromycological study in the Cathedral of Santiago de Compostela (Spain). *Int Biodeterior Biodegrad* 60(4):231–237.

Ascione, F., L. Bellia, A. Capozzoli, and F. Minichiello. 2009. Energy saving strategies in air-conditioning for museums. *Appl Therm Eng* 29(4):676–686.

Barbeau, D. N., L. F. Grimsley, and L. E. Whitte. 2010. Mold exposure and health effects following Hurricanes Katrina and Rita. *Annu Rev Public Health* 31:165–178.

Beckett, W., M. Kallay, A. Sood, Z. Zuo, and D. Milton. 2005. Hypersensitivity pneumonitis associated with environmental Mycobacteria. *Environ Health Perspect* 113(5):767–767.

Bernstein, D. I., Z. L. Lummus, G. Santilli, J. Siskosky, and I. L. Bernstein. 1995. Machine operator's lung: A hypersensitivity pneumonitis disorder associated with exposure to metalworking fluid aerosols. *Chest* 108(3):636–641.

Bonetta, S., S. Mosso, S. Sampo, and E. Carraro. 2010. Assessment of microbiological indoor air quality in an Italian office building equipped with an HVAC system. *Environ Monit Assess* 161:437–483.

Borrego, S., P. Lavin, I. Perdomo, S. Gómez de Saravia, and P. Guiamet. 2012. Determination of indoor air quality in archives and biodeterioration of the documentary heritage. *ISRN Microbiol* 2012:680598.

Brickus, L. S. R., L. F. G. Siqueira, F. R. Aquinto Neto, and J. N. Cardoso. 1998. Occurrence of airborne bacteria and fungi in bayside offices in Rio de Janeiro, Brazil. *Indoor Built Environ* 7(5–6):270–275.

Busato, A., D. Hofer, T. Lentze, C. Gaillard, and A. Burnens. 1999. Prevalence and infection risks of zoonotic enteropathogenic bacteria in Swiss cow-calf farms. *Vet Microbiol* 69(4):251–263.

Carducci, A., M. Verani, and R. Battistini. 2010. Legionella in industrial cooling towers: Monitoring and control strategies. *Lett Appl Microbiol* 50(1):24–29.

Cole, E. C., K. K. Foarde, K. E. Leese, D. A. Green, D. L. Franke, and M. A. Berry. 1993. Biocontaminants in carpeted environments. In: *Proceedings of the 6th International Conference on Indoor Air Quality and Climate, Indoor Air '93*, Helsinki, Finland, vol. 4, 351–356.

Coron, S., and M. Lefevre. 1993. Lutte contre les moisissures l'experience de la Bibliotheque de l'Arsenal. *Bulletin des bibliothèques Bibl de France* 4:104–109.

Cox, C. S., and C. M. Wathes, eds. 1995. *Bioaerosols Handbook*. Boca Raton, FL: Lewis Publishers/CRC Press, Inc.

Crook, B., and N. C. Burton. 2010. Indoor moulds, sick building syndrome and building related illness. *Fungal Biol Rev* 24(3–4):106–113.

Cyprowski, M., and J. A. Krajewski. 2003. Czynniki szkodliwe dla zdrowia występujące w oczyszczalniach ścieków komunalnych. [Health hazards in municipal wastewater treatment plants]. *Med Pr* 54(1):73–80.

Cyprowski, M., A. Ławniczek-Wałczyk, and R. L. Górny. 2016. Airborne peptidoglycons as a supporting indicator of bacterial contamination in a metal processing plant. *Int J Occupation Med Environ Health* 29(3):427–437.

Darragh, A. H., R. M. Buchan, D. R. Sandford, and R. O. Coleman. 1997. Quantification of air contaminants at a municipal sewage sludge composting facility. *Appl Occup Environ Hyg* 12(3):190–194.

Desjardins, E., and C. Beaulieu. 2003. Identification of bacteria contaminating pulp and a paper machine in a Canadian paper mill. *J Ind Microbiol Biotechnol* 30(3):141–145.

Doleżal, M. 1993. Zanieczyszczenie mikrobiologiczne ze szczególnym uwzględnieniem grzybów w budownictwie mieszkaniowym [Microbiological contamination with particular emphasis on fungi in residential buildings]. In: *Materiały Krajowej Konferencji "Zdrowy Dom", 23–24 września 1993*, Warsaw, 185–189.

Dutkiewicz, J. 1989. Bacteria, fungi and endotoxin in stored timber logs and airborne saw dust in Poland. *Biodeterior Res* 2:533–547.

Dutkiewicz, J. 1997. Bacteria and fungi in organic dust as potential health hazard. *Ann Agric Environ Med* 4:11–16.

Dutkiewicz, J., W. G. Sorenson, D. M. Lewis, and S. A. Olenchock. 1992. Levels of bacteria, fungi and endotoxin in stored timber. *Internat Biodeterior* 30(1):29–46.

Dutkiewicz, J., E. Krysińska-Traczyk, Z. Prażmo, C. Skořska, and J. Sitkowska. 2001a. Exposure to airborne microorganisms in Polish sawmills. *Ann Agric Environ Med* 8(1):71–80.

Dutkiewicz, J., S. Olenchock, E. Krysińska-Traczyk, C. Skórska, J. Sitkowska, and Z. Prazmo. 2001b. Exposure to airborne microorganisms in fiberboard and chipboard factories. *Ann Agric Environ Med* 8(2):191–199.

El-Nagerabi, S. A. F., A. E. Elshafie, and U. A. Al-Hinai. 2014. The mycobiota associated with paper archives and their potential control. *Nus Biosci* 6:19–25.

Farhat, M., R. A. Shaheed, and H. H. Al-Ali 2018. Legionella confirmation in cooling tower water. Comparison of culture, real-time PCR and next generation sequencing. *Saudi Med J* 39(2):137–141.

Fazio, A. T., L. Papinutti, B. A. Gómez et al. 2010. Fungal deterioration of a Jesuit South American polychrome wood sculpture. *Int Biodeterior Biodegrad* 64(8):694–701.

Flannigan, B., E. M. McCabe, and F. McGarry. 1991. Allergenic and toxigenic micro-organisms in houses. *J Appl Bacteriol, Symp Suppl* 70:61S–73S.

Flemming, H. C., M. Meier, and T. Schild. 2013. Mini-review: Microbial problems in paper production. *Biofouling* 29(6):683–696.

Gołofit-Szymczak, M., and R. L. Górny. 2010. Bacterial and fungal aerosols in air-conditioned office buildings in Warsaw, Poland: The winter season. *Int J Occup Saf Ergon* 16(4):465–476.

Górny, R. L., M. Gołofit-Szymczak, M. Cyprowski, and A. Stobnicka-Kupiec. 2019. Nasal lavage as analytical tool in assessment of exposure to particulate and microbial aerosols in wood pellet production facilities. *Sci Total Environ* 20(697).

Gots, R. E., N. J. Layton, and S. W. Pirages. 2003. Indoor health: Background levels of fungi. *Am Ind Hyg Assoc J* 64(4):427–438.

Guglielminetti, M., C. De Giuli Morghen, A. Radaelli et al. 1994. Mycological and ultrastructural studies to evaluate biodeterioration of mural paintings: Detection of fungi and mites in Frescos of the monastery of St Damian in Assisi. *Int Biodeterior Biodegrad* 33(3):269–283.

Haleem Khan, A. A., and S. Mohan Karuppayil. 2012. Fungal pollution of indoor environments and its management. *Saudi J Biol Sci* 19(4):405–426.

Hempel, M., V. Rakhra, A. Rothwell, and D. Song. 2014. Bacterial and fungal contamination in the library setting: A growing concern? *Environ Health Rev* 57(1):9–15.

Herbarth, O., U. Schlink, A. Müller, and M. Richter. 2003. Spatiotemporal distribution of airborne mould spores in apartments. *Mycol Res* 107(11):1361–1371.

Hung, L. L., J. D. Miller, and K. Dillon, eds. 2005. *Field Guide for the Determination of Biological Contaminants in Environmental Samples*. Fairfax, VA: AIHA.

Jager, E., and C. Eckrich. 1997. Hygienic aspects of biowaste composting. *Ann Agric Environ Med* 4:99–105.

Joblin, Y., S. Moularat, R. Anton et al. 2010. Detection of moulds by volatile organic compounds: Application to heritage conservation. *Int Biodeterior Biodegrad* 64(3):210–217.

Kowalski, W. J., ed. 2006. *Aerobiological Engineering Handbook: A Guide to Airborne Disease Control Technologies*. New York: The McGraw-Hill.

Kozajda, A., M. Sowiak, and M. Piotrowska. 2009. Sortownia odpadów komunalnych: Rozpoznanie narażenia na czynniki biologiczne (grzyby strzępkowe) [Municipal waste sorting plant: Recognition of exposure to biological agents (filamentous fungi)]. *Med Pr* 60(6):483–490.

La Rosa, G., M. Fratini, S. Della Libera, M. Iaconelli, and M. Muscillo. 2013. Viral infections acquired indoors through airborne, droplet or contact transmission. *Ann Ist Super Sanità* 49(2):124–132.

Laitinen, S. K. 1999. Importance of sampling, extraction and preservation for the quantitation of biologically active endotoxin. *Ann Agric Environ Med* 6(1):33–38.

Ławniczek-Wałczyk, A., M. Gołofit-Szymczak, M. Cyprowski, and R. L. Górny. 2012. Exposure to harmful microbiological agents during the handling of biomass for power production purposes. *Med Pr* 63(4):395–407.

Ławniczek-Wałczyk, A., R. L. Górny, M. Gołofit-Szymczak, A. Niesler, and A. Wlazło. 2013. Occupational exposure to airborne microorganisms, endotoxins and β-glucans in poultry houses at different stages of the production cycle. *Ann Agric Environ Med* 20(2):259–268.

Lewis, D. M., J. Dutkiewicz, W. G. Sorenson, M. Mamolen, and J. E. Hall. 1990. Microbiological and serological studies of an outbreak of "humidifier fever" in a print shop. *Biodeterior Res* 3:467–477.

Mackiewicz, B., J. Dutkiewicz, J. Siwiec et al. 2019. Acute hypersensitivity pneumonitis in woodworkers caused by inhalation of birch dust contaminated with *Pantoea agglomerans* and *Microbacterium barkeri*. *Ann Agric Environ Med* 26(4):644–655.

Maggi, O., A. M. Persiani, F. Gallo et al. 2000. Airborne fungal spores in dust present in archives: Proposal for a detection method, new for archival materials. *Aerobiologia* 16(3/4):429–434.

Mandrioli, P., G. Caneva, and C. Sabbioni. 2003. *Cultural Heritage and Aerobiology: Methods and Measurement Techniques for Biodeterioration Monitoring*. Dordrecht: Kluwer Academic Publishers.

Martin, W. T., Y. Zhang, P. Willson, T. P. Archer, C. Kinahan, and E. M. Barber. 1996. Bacterial and fungal flora of dust deposits in a pig building. *Occup Environ Med* 53(7):484–487.

Masclaux, F. G., O. Sakwinska, N. Charrière, E. Semaani, and A. Oppliger. 2013. Concentration of airborne Staphylococcus aureus (MRSA and MSSA), total bacteria, and endotoxins in pig farms. *Ann Occup Hyg* 57(5):550–557.

Mc Donnell, P. E., M. A. Coggins, V. J. Hogan, and G. T. Fleming. 2008. Exposure assessments of airborne contaminants in the indoor environment of Iris swine farms. *Ann Agric Environ Med* 15(2):323–326.

Mesquita, N., A. Portugal, S. Videira et al. 2009. Fungal diversity in ancient documents: A case study on the Archive of the University of Coimbra. *Int Biodeterior Biodegrad* 63(5):626–629.

Midtgaard, U., and O. M. Poulsen. 1997. Waste collection and recycling: Bioaerosol exposure and health problems. *Ann Agric Environ Med* 4:1–176.

Morey, P. R., and J. C. Feeley. 1990. The landlord, tenant, and investigator: Their needs, concerns, and viewpoints. In: *Biological Contaminants in Indoor Environments*, eds. P. R. Morey, J. C. Feeley, and J. A. Otten, 1–20. Philadelphia, PA: American Society for Testing and Materials.

Nevalainen, A. 1989. *Bacterial Aerosols in Indoor Air.* Helsinki: National Public Health Institute.

Patterson, W. B., D. E. Craven, D. A. Schwartz, E. A. Nardell, J. Kasmer, and J. Noble. 1985. Occupational Hazards to hospital personnel. *Ann Intern Med* 102(5):658–680.

Petrycka, H. 1993. *Ćwiczenia z mikrobiologii środowiskowej.* Gliwice: Wydawnictwo Politechniki Śląskiej.

Pinheiro, A. C., M. F. Macedo, J. C. Saiz-Jimenez et al. 2011. Mould and yeast identification in archival settings: Preliminary results on the use of traditional methods and molecular biology options in Portuguese archives. *Int Biodeterior Biodegrad* 65(4):619–627.

Pope, A. M., R. Patterson, and H. Burge, eds. 1993. *Indoor Allergens: Assessing and Controlling Adverse Health Effects.* Washington, DC: National Academy Press.

Popescu, S., C. Borda, E. A. Diugan, and D. Oros. 2014. Microbial air contamination in indoor and outdoor environment of pig farms. *Animal Sci. Biotech* 47:182–187.

Prażmo, Z., J. Dutkiewicz, and G. Cholewa. 2000. Gram-negative bacteria associated with timber as a potential respiratory hazard for woodworkers. *Aerobiologia* 16(2):275–279.

Prażmo, Z., J. Dutkiewicz, C. Skorska, J. Sitkowska, and G. Cholewa. 2003. Exposure to airborne Gram-negative bacteria, dust and endotoxin in paper factories. *Ann Agric Environ Med* 10(1):93–100.

Prussin, A. J. II, E. B. Garcia, and L. C. Marr. 2015. Total virus and bacteria concentrations in indoor and outdoor air. *Environ Sci Technol Lett* 2(4):84–88.

Rao, C. Y., M. A. Riggs, G. L. Chew et al. 2007. Characterization of airborne molds, endotoxins, and glucans in homes in New Orleans after Hurricanes Katrina and Rita. *Appl Environ Microbiol* 73(5):1630–1634.

Riggs, M. A., C. Y. Rao, C. M. Brown et al. 2008. Resident cleanup activities, characteristics of flood-damaged homes and airborne microbial concentrations in New Orleans, Louisiana, October 2005. *Environ Res* 106(3):401–409.

Rylander, R., and Y. Peterson. 1994. Causative agents for organic dust related disease. *Am J Ind Med* 25(1):1–146.

Seedorf, J. 2004. An emission inventory of livestock-related bioaerosols for Lower Saxony, Germany. *Atm Environ* 38(38):6565–6581.

Skóra, J., B. Gutarowska, K. Pielech-Przybylska et al. 2015. Assessment of microbiological contamination in the work environments of museums, archives and libraries. *Aerobiologia* 31(3):389–401.

Sterflinger, K. 2010. Fungi: Their role in deterioration of cultural heritage. *Fungal Biol Rev* 24(1–2):47–55.

Strzelczyk, A. B., and J. Karbowska-Berent. 2004. *Drobnoustroje i owady niszczące zabytki i ich zwalczanie* [Microorganisms and Insects Destroying Monuments and Fighting Them]. Torun: Wydawnictwo Uniwersytetu Mikołaja Kopernika.

Sundell, J., H. Levin, W. W. Nazaroff et al. 2011. Ventilation rates and health: Multidisciplinary review of the scientific literature. *Indoor Air* 21(3):191–204.

Szadkowska-Stańczycz, I., ed. 2007. *Zagrożenia i skutki zdrowotne narazenia na szkodliwe czynniki biologiczne pracowników zakładów gospodarki odpadami* [Hazards and Health Effects of Exposure to Biological Agents of Workers in Waste Management Plants]. Łódź: Instytut Medycyny Pracy im. Prof. J. Nofera.

Szadkowska-Stańczyk, I., K. Bródka, A. Buczyńska, M. Ceprowski, A. Kozajda, and M. Sowiak. 2010. Exposure to bioaerosols among cafo workers (swine feeding). *Med Pr* 61(3):257–269.

Tillie-Leblond, I., F. Grenouillet, G. Reboux et al. 2011. Hypersensitivity pneumonitis and metalworking fluids contaminated by mycobacteria. *Eur Respir J* 37(3):640–647.

Tseng, C. H., H. C. Wang, N. Y. Xiao, and Y. M. Chang. 2011. Examining the feasibility of prediction models by monitoring data and management data for bioaerosols inside office buildings. *Build Environ* 46(12):2578–2589.

Turetgen, I., and A. Cotuk. 2007. Monitoring of biofilm-associated L. pneumophila on different substrata in model cooling tower system. *Environ Monit Assess* 125(1–3):271–279.

Valentin, N. 2003. Microbial contamination and insect infestation in organic materials. *Coalition* 6:2–5.

Zielińska-Jankiewicz, K., A. Kozajda, M. Piotrowska, and I. Szadkowska-Stańczyk. 2008. Microbiological contamination with moulds in work environment in libraries and archive storage facilities. *Ann Agric Environ Med* 15(1):71–78.

Zuskin, E., J. Mustajbegovic, E. N. Schachter et al. 1995. Respiratory function in poultry workers and pharmacologic characterization of poultry dust extract. *Environ Res* 70(1):11–19.

Zyska, B. 1997. Fungi isolated from library materials: A review of the literature. *Int Biodeterior Biodegrad* 40(1):43–51.

# Index

Printed in the United States
by Baker & Taylor Publisher Services